D0919639

FUNDAMENTALS OF SPORTS BIOMECHANICS

FUNDAMENTALS OF SPORTS BIOMECHANICS

CHARLES SIMONIAN

Ohio State University

PRENTICE-HALL, INC.
Englewood Cliffs, New Jersey 07632

Library of Congress Cataloging in Publication Data

Simonian, Charles. (date)
Fundamentals of sports biomechanics.

Bibliography: p.
Includes index.
1. Sports—Physiological aspects. 2. Human mechanics. I. Title.
RC1235.S54 612'.76 80-29214
ISBN 0-13-344499-6

Printed in the United States of America

10 9 8 7 6 5 4 3 2 1

Editorial/production supervision
and interior design by Barbara Alexander
Cover design by Wanda Lubelska
Manufacturing buyer: Harry P. Baisley

PRENTICE-HALL INTERNATIONAL, INC., *London*
PRENTICE-HALL OF AUSTRALIA PTY. LIMITED, *Sydney*
PRENTICE-HALL OF CANADA, LTD., *Toronto*
PRENTICE-HALL OF INDIA PRIVATE LIMITED, *New Delhi*
PRENTICE-HALL OF JAPAN, INC., *Tokyo*
PRENTICE-HALL OF SOUTHEAST ASIA PTE. LTD., *Singapore*
WHITEHALL BOOKS LIMITED, *Wellington, New Zealand*

To my wife, Carol
and to my children,
Rick, Julia, Karen, and Laura

Contents

CHAPTER 3

Force 29

CHAPTER 4

Linear motion 57

CHAPTER 5

Angular motion 97

Preface

Physical education teaching and athletic coaching have undergone many changes in recent years. Teachers are better educated than ever before, continue to show concern for the welfare of their pupils and athletes, and are more in tune with current research and sports medicine. One of the many reasons for this progress has been the increased emphasis on the science of movement in the physical education professional preparation programs in our colleges and universities. The core science courses are exercise physiology, motor learning, kinesiology and biomechanics.

Kinesiology, the study of human movement, has long been a required subject for students majoring in physical education, but it generally was a single course that emphasized either gross or functional anatomy. The mechanical bases of movement, previously given perfunctory treatment, began receiving ever-increasing attention, so much so that many instructors had difficulty covering the necessary mechanics and anatomy in a single course. At some institutions, the solution has been to divide kinesiology into two courses, one devoted to functional anatomy and sometimes labeled *anatomical kinesiology,* and the other termed *mechanical kinesiology,* or, as is now more in vogue, *biomechanics.* Whatever the titles used, an understanding of both areas is essential for physical-educators, coaches, dance teachers, and others in the business of teaching or analyzing human motion.

The subject of biomechanics either can be treated rather generally, almost superficially, and applied to sports situations, or can be dealt with on the highly sophisticated, quantitative level necessary in research. This text admits to the generalized, qualitative approach and attempts to be both practical and useful to teachers and coaches—the practitioners—while also providing a first course of study for those who would go further into this most difficult and interesting field.

This book is addressed to a varied audience. Undergraduate students of physical education, athletic coaching, and dance should find the material relevant to their professional needs in kinesiology, and they may be stimulated to make their own applications of the principles included herein to their particular sports interests.

Practicing teachers and coaches who at some past time have taken kinesiology courses limited to functional anatomy may now feel a need to understand the mechanics underlying many of the skills they have been teaching. Admittedly, a knowledge of the proper sports techniques (the *how* of an activity) and of the best means to convey the knowledge (the pedagogy) is much more important, on the day-to-day level, than is the knowledge of the science of movement (the *why*). But knowing the *why* makes the difference between curious, thorough, professional educators and unquestioning robots. The former have confidence in what they teach, keep abreast of developments, and can justify their programs, while the latter are satisfied that what they have been teaching works well enough and there is no need to change.

For the graduate student of physical education, especially one who plans someday to teach kinesiology to undergraduates, a knowledge of biomechanics is obviously a necessary part of his or her education. Such students would be well advised to take also one or more courses in physics and mathematics.

Finally, this book can serve as an introduction to biomechanics for those few graduate students who intend to specialize and do research in this field. Such students must be prepared to obtain the near equivalent of a degree in physics or engineering mechanics and must have a solid background in mathematics, anatomy, physiology, cinematography, computer science, and perhaps electromyography.

It is hoped that the study of this text will increase the reader's awareness of the scope of the field and will provide some capacity to comprehend the increasing amount of biomechanical research being reported in the journals and being applied toward the improvement of sports performance.

No prerequisite knowledge of physics or mathematics beyond what is ordinarily taken in high school is necessary to understand the material in this book. Clearly, those students who are well prepared in those areas will have a decided advantage over those who are not, but the necessary mathematics and physical principles are presented in a svstematic and simple manner that should be understandable to all.

Unlike a number of other sports biomechanics texts, this one makes no attempt to be all-inclusive, and despite the inherent dangers of making such value judgments, a number of topics normally found in biomechanics or kinesiology texts are either treated lightly or purposely omitted because they were felt to have too little utility in sports or to be too complex to deal with at this level. Such omissions are of course arbitrary, and the author does not wish to imply that all knowledge must have an immediate practical use; but a number of excellent books exist for those students who may require broader, in-depth material. A number of these are cited in the references.

The order of topic presentation was difficult to decide, because so many principles overlap and so many depend upon a prior understanding of others that almost no sequence is entirely satisfactory. Even the simplest movement skills involve several principles and can be described from several viewpoints. It was decided to arrange the material in a logical learning progression rather than to group the principles under such headings as kinetics and kinematics.

Where possible, the topics have been written in a format that presents mechanical principles and follows these with the related mathematical concepts and some worked examples. Some applications are included in each chapter. Graduate students should be expected to absorb all of the material, while undergraduates can obtain a reasonable understanding of biomechanics by studying only the nonmathematical portions. It should be stressed, however, that the mathematics used here is relatively simple and is considered a valuable reinforcement in the learning and comprehension of the principles. This attitude is reflected by MacConaill and Basmajian, who state, "Mathematics is indispensable if kinesiology is to be more than a set of statements learnt by rote, if it is to be understood in such a way that it can both explain what we do know about movements and muscles and also form the basis of further, fruitful inquiry."[1]

I would like to express thanks to the many Ohio State University students who provided valuable critiques of early drafts of the text which were used in class and to Dr. Louis E. Alley of the University of Iowa, who aroused my first interest in biomechanics.

[1] M. A. MacConnaill and J. V. Basmajian, *Muscles and Movements* (Baltimore: Williams & Wilkins Company, 1969), p. v.

FUNDAMENTALS OF SPORTS BIOMECHANICS

CHAPTER 1

Introduction

Those who are interested in the scientific aspects of sports and physical education have known for some time that it is not enough to try to explain movement simply in terms of musculoskeletal activity. It has been found to be increasingly necessary to make appropriate applications of certain aspects of physics to the study of human motion, and so traditional kinesiology courses over recent years have devoted more and more time to the mechanics of movement. This trend has brought the emergence of a number of pioneer scholars in a specialized field of study commonly called *biomechanics.* But this title does not suffice for those of us in the areas of physical education and athletics, because it already covers too large a range of interests, some of them not directly related to sports. Biomechanics is and has been important in the design of space vehicles, automobiles, and aircraft; in industrial contexts; and in medicine and rehabilitation. An examination of a library card catalog under the subject heading *biomechanics* will disclose that this title encompasses every imaginable relationship between living tissue and engineering mechanics. Therefore, it may be more descriptive of our special interests to call the study *sports biomechanics.*

This is not to say that athletic performance should be analyzed purely in mechanical terms. A great many physiological and psychological factors can and do affect human learning and sports performance, and these relationships, what little we presently know of them, are important studies in themselves. It has become amply clear that there will have to be an integration of knowledge from many disciplines before any real progress can be expected in the understanding of human movement. Still, biomechanical laws can be related to all human motion, and a working knowledge of these laws should be possessed by all physical educators and coaches.

A background in mechanics can help coaches to know their sports more fully, can make them more confident about their practices, and can extend their knowledge beyond the techniques involved in the sports to the underlying scientific reasons for doing a particular movement in a given way. They will be better prepared to answer pupils' questions such as *Coach, why do it that way?* and *But coach, why shouldn't I do it this way?* The student deserves a better answer than *Because I said so!* Those coaches who teach only as they themselves were taught, never questioning the reasons behind their methods, are limiting their total effectiveness as teachers of young athletes.

While no data are available for all sports, it is probable that less than half of all the coaches in this country have had formal physical education training or have been exposed to a course in kinesiology or biomechanics. Included in this estimate of those not so trained are thousands of lay coaches of youngsters in sports conducted outside the public schools. These noncertified coaches, many of them parents whose children are in the sport, are rendering a vital service to our youth, and they deserve more attention than they are getting from professional educators. Clinics on exercise physiology, motor learning, kinesiology, biomechanics, and athletic training should be provided within every community.

The complete coach, then, in addition to being reasonably skilled in his or her own sport, should know something about pedagogy, anatomy, exercise physiology, sports psychology, and mechanics. Only by having such competencies can the coach make the most of the developing athletes.

Readers of this text must be cautioned that they will encounter considerable difficulty if they try to apply laws of physics with great precision to human movement. The human body is neither a particle nor a rigid body, either of which is relatively easy to study, but is, rather, a flexible, dynamic mass of many segments, capable of performing many different and simultaneous movements. For this reason, it is unusual for a researcher to be able to conduct even simple experiments on human motion in which all conditions are adequately controlled. Still, a good many scholars in sports biomechanics are striving for and making progress toward excellence in measurement and methodology. Computers and newly developed instrumentation, well described by Miller and Nelson,[1] have expedited previously laborious hand data processing and have opened new avenues for investigation.

While one must be extremely careful about making generalizations or loosely applying principles of mechanics to sports situations, some useful generalizations can be made, and at this early stage of the science of sports

[1] Doris I. Miller and R. C. Nelson, *Biomechanics of Sport* (Philadelphia: Lea & Febiger, 1973).

biomechanics, we may be justified in taking some liberties, stretching some laws to fit, and making some assumptions that would not be acceptable to a physicist or an engineer who deals with rigid bodies and mass particles.

Biomechanical research can make use of the accepted, classical techniques used in physics and statistics, or the research can be of an applied nature. Either approach may require complex instrumentation and an extensively trained scientist, but applied research does not preclude a degree of empirical reasoning by experienced coaches and athletes seeking to improve performance but uninterested in the development of theory or in the use of mathematical equations.

It was this kind of thinking, perhaps along with some trial and error, that brought about dramatic style changes in, for example, shot-putting and the high jump. Athletes and coaches who have been willing to tinker, for want of a better word, with traditional techniques, training regimens, and stratagems probably have accounted for most of the improvements in sports performances. We would be hard pressed to find evidence that such improvement is a direct result of laboratory research. This may be because biomechanical research has been preoccupied with studying the what and the why of a superior performance or style and has been notably negligent in the area of invention of better techniques. As Higgins states, "All too often the approach has been to focus on highly skilled performances and then attempt directly to apply what has been learned to teaching and coaching at all levels of skill development."[2] There is no doubt that this situation is changing as the isolated efforts of scientists are being shared with others through writing and presentations at professional meetings.

The First International Seminar on Biomechanics was held in Zurich in 1967, and there has been a seminar every two years since that time. The proceedings of each of these sessions have been published and provide a valuable resource for biomechanics students. There have also been a number of biomechanics conferences held on specific sports themes, such as swimming.

Excellence in sports has long been of political value, especially in Eastern European countries, where many governments have provided financial support to the various sports sciences, including biomechanics. Results of such concentrated efforts are evident today in international competitions, which are increasingly being dominated by the nations that have the most advanced sports science facilities.

But biomechanics offers no miracles or guarantees of instant success for coaches who take up its study. In fact, there are many instances in sports where the technique being used is not dictated by mechanical laws so much as

[2] Joseph R. Higgins, *Human Movement: An Integrated Approach* (Saint Louis: C. V. Mosby Company, 1977), p. 3.

by rules, safety and esthetic considerations, available equipment, and human structural and physiological limitations. Good and bad performances in many sports are not determined so much by biomechanical technique as by strategic sense, courage to execute, and judgment of time and distance. All of these are directly related to experience and training.

The science of sports biomechanics is young, and as more researchers are prepared, as technology is improved, as more facilities are developed, and as more funds become available from public and private sources, we can certainly expect practical outcomes on the athletic fields. Perhaps then the invention of new techniques will occur in the laboratory before they are discovered on the field by the athlete through chance or experimentation.

BASIC TERMS

Mechanics is that branch of science which studies the motion or form of bodies under the action of forces.

Biomechanics is the field of study which applies the principles of mechanics to the structure and movement of living things.

Sports biomechanics is the application of the principles of biomechanics to the study of human motion in sports and exercise.

Dynamics is that branch of mechanics dealing specifically with unbalanced forces which produce some change in the state of motion of a body.

Kinematics is a subdivision of dynamics which is concerned with quantitative descriptions, such as of an object's position in space, velocity, and acceleration, without regard for the forces involved.

Kinetics is a subdivision of dynamics which is concerned with the effects of unbalanced forces, that is, with the causes of motion.

Statics is that branch of mechanics dealing specifically with balanced forces which produce no change in the state of motion of a body. The focus is on bodies which are unaccelerated and therefore in equilibrium.

As an example of how these fields of study apply to human movement, let us make a general analysis of a speed-swimming start. In Figure 1–1, when the swimmer is holding his position awaiting the firing of the starting gun, he is in equilibrium, and an analysis of his motionless position comes under *statics* (Figure 1–1*a*). When the gun sounds, the swimmer applies muscular force to dive from the block, and as he dives, the force of gravity

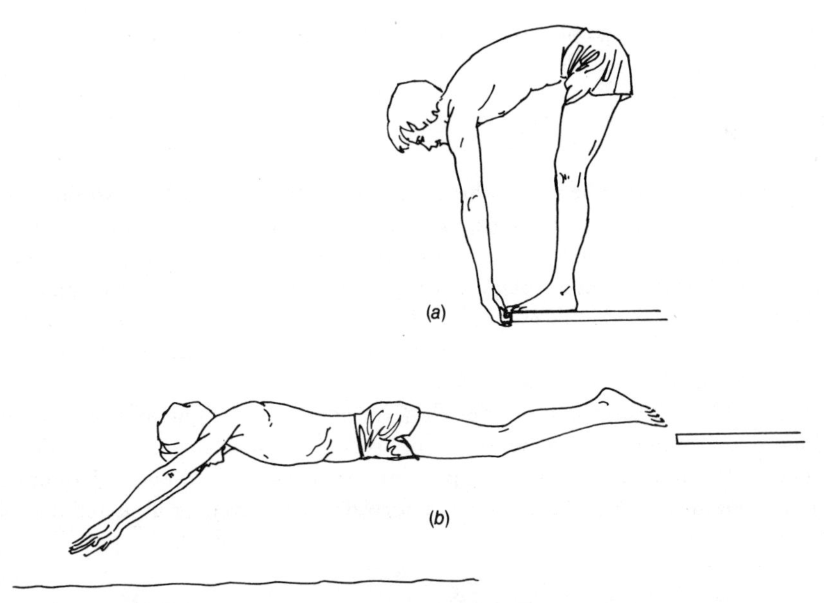

Figure 1-1 (*a*) A swimmer in the blocks is a study in statics. (*b*) During takeoff, the swimmer's acceleration, velocity, angle of entry, and flight path are the concerns of kinematics, and the forces that cause the changes in motion are the concerns of kinetics.

pulls him down to the water (Figure 1-1*b*). These forces causing his motion forward and downward are the concerns of *kinetics. Kinematics* deals with the swimmer's pattern of flight or with the speed with which he leaves the block or enters the water.

DESCRIBING MOVEMENT

The task of describing human motion is made simpler if a basic frame of reference, a common language, is employed that is understood by other biomechanicians or kinesiologists. Over the years, effective use has been made of such standard anatomic terms as *flexion, extension, abduction, adduction, proximal,* and *distal,* and anyone who deals with human movement should be thoroughly familiar with these. Appendix A provides a synopsis of anatomic terms. While they are adequately descriptive of movements of single body segments, these terms are not always easily applied to complex whole-body movements in sports.

Most segmental human movement is angular in nature; that is, bones move around joints, and such motion takes place in some plane around some axis.

PLANES

In common usage, there are three classes of planes: sagittal, lateral, and transverse. One plane of each class that passes through the human body's center of gravity is referred to as a *primary* plane, and all other planes which are parallel to the primary plane but which do not pass through the center of gravity are known as *secondary* planes. The three primary planes all intersect at right angles at the center of gravity of a person standing in the anatomic position, that is, in an erect stance with palms facing forward.

The *sagittal* plane divides the erect body into right and left halves (Figure 1-2). All flexion, extension, and hyperextension movements made by a person standing in the anatomic position are in the sagittal plane. Examples of movements made in this plane are a forward or backward roll, a nodding of

Figure 1-2 The sagittal plane and lateral axis

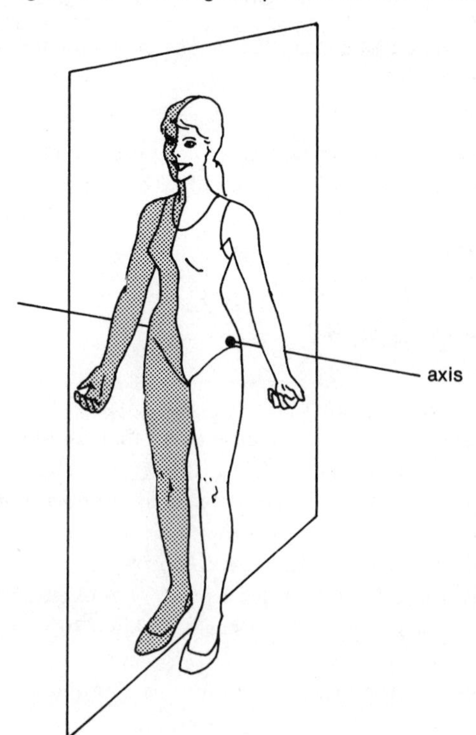

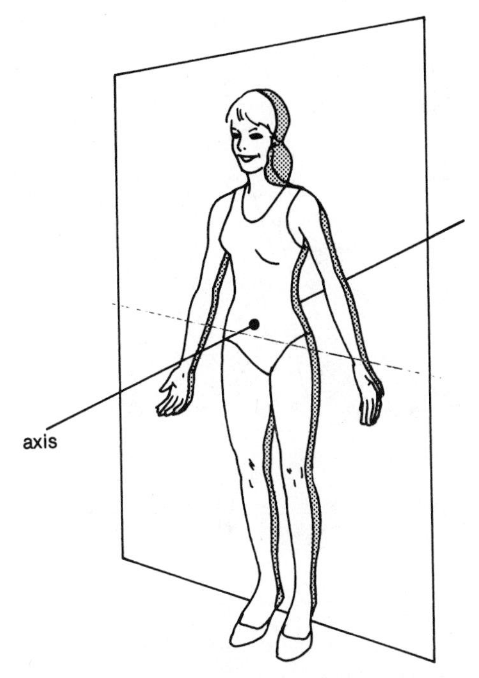

Figure 1-3 The lateral plane and sagittal axis

the head, a punt, a bow from the waist, running, a backbend, and a barbell curl.

The *lateral* plane divides the erect human body into front and rear halves (Figure 1-3). All abduction and adduction movements of the extremities and lateral flexions of the spine occur in this plane. Representative movements in the lateral plane are cartwheels, straddle hops, and side bends.

The *transverse* plane divides the erect body into equal-mass sections above and below the center of gravity. Such rotational movements as a jump-turn, inward rotation of the humerus, and turning of the head left or right are all done in this plane (Figure 1-4).

AXES

Angular motion is defined in Chapter 5 as motion around an axis. All of the movements that were just described are of an angular nature and so must take place around some axis which is perpendicular to the particular plane. When the plane is primary, it and the axis for the movement will pass through the center of gravity of the body. This is best illustrated by the rotation of air-

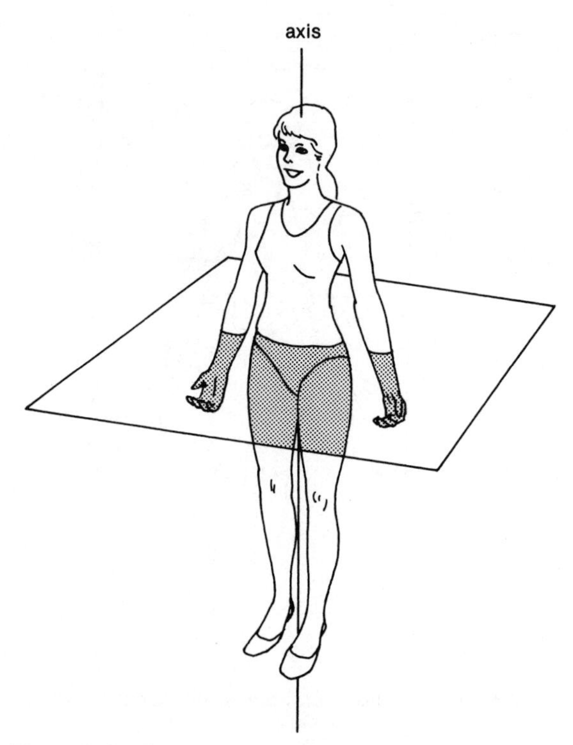

Figure 1-4 The transverse plane and longitudinal axis

borne bodies, as such rotations always occur around an axis which passes through the mass center. When the plane is secondary, the axis will be found passing through the center of the joint involved.

Movements made in a sagittal plane occur around a *lateral* axis (Figure 1–2). Those made in a lateral plane turn around a *sagittal* axis (Figure 1–3). Movements which are in the transverse plane will have a *longitudinal* axis (Figure 1–4).

It should be noted that the names given here for the planes and axes, while fairly common, are not standardized, and the reader will find various other names used in other texts. Some frequently used synonyms are:

For *sagittal: antero-posterior*
For *lateral: frontal, coronal*
For *transverse: horizontal*
For *longitudinal: vertical*

When determining the plane and axis for any movement that is being analyzed, imagine the subject to be standing in the anatomic position.

Then, regardless of the performer's actual position in space, the plane and axis for the whole body or for any segment can be more readily identified.

A gymnast doing giant swings on a horizontal bar is moving in a sagittal plane around a lateral axis, the bar. If the gymnast in Figure 1-5 does a somersault dismount, he will continue to rotate in the sagittal plane, but his lateral axis changes from the bar to an axis which now passes through his center of gravity. This is because the axis for rotation of any object in space must pass through the object's center of gravity.

In delivering a ball, a bowler's arm moves in a secondary sagittal plane around a lateral axis which passes through the center of the shoulder joint. Until it is released, the ball may be considered a fixed part of the arm, but after release it will move in its own plane and around a new axis of rotation.

These examples are admittedly rather easy to analyze because the plane is obvious. In practicing the analysis of more complex sports movements, the student will soon detect that this system has some severe limitations and cannot easily be used to describe all of the vast range of possible movements. But it is a starting point and does have some value in motion analysis and in the communication of information.

Figure 1-5 While the gymnast is swinging, the bar serves as the axis for rotation, but during the somersault dismount, the gymnast rotates around his center of gravity.

CENTER OF GRAVITY

The center of gravity is the point at which all the weight or mass of a body may be considered to be concentrated. As noted earlier, the center of gravity of an individual standing in the anatomic position marks the intersection of the three primary planes and their axes. For solid masses of uniform density, the location of this point, often called the *mass center,* is at the geometric center and remains constant no matter what position the object assumes. In rings or hoops, the center of gravity is found in the air space at the center of the circle.

The human body's flexibility and its fluid internal structure create great problems in accurately locating the center of gravity, because, while the mass center can be determined for any given, momentarily fixed stance, any major movement is accompanied by a shift in the location of the center of gravity. Thus, in many sports skills the mass center is constantly moving.

Locating the mass center of a rigid object is not difficult and is even easier if the object is of uniform density and of a symmetrical shape, in which case the center of gravity is at the exact center of the object. An object suspended from this point is in *rotational equilibrium.*

Probably the simplest demonstration of center of gravity determination involves the use of any rigid, flat, irregularly shaped board material which has several randomly drilled holes around its perimeter. Suspend both the board and a plumb line from a nail, and use chalk to trace the plumb line on the board. Rotate the board to suspend it from another hole and draw a line along the path of the plumb. Repeat the procedure once more and note where the drawn lines intersect on the board (Figure 1-6). That point of intersection should mark the center of gravity. If the board were to be tossed up into the air with some spin in the sagittal plane, the board would be seen to rotate around the point marked as the mass center. If the board were to be tossed in a forward and upward direction, whether spinning or not, its mass center would follow a parabolic path. In other words, it would behave as would a particle or a point with no structure to consider. Similarly, as will be discussed in Chapter 4, the path of any airborne athlete's center of gravity describes a parabola, and no man or woman's physical makeup or muscular actions in the air will have an effect upon the path of the "particle" called the body's center of gravity.

Since the late nineteenth century, there have been a number of efforts to locate the center of gravity of the human body and of the body's various segments. The methods used have varied from measuring frozen cadavers, as was done by Dempster,[3] to the use of templates designed by Walton.[4] Some researchers have employed immersion techniques for obtain-

[3] W. T. Dempster, "Space Requirements of the Seated Operator," WADC Technical Report 55-159 (Wright-Patterson Air Force Base, Ohio: 1955).

[4] J. S. Walton, "A Template for Locating Segmental Centers of Gravity," *Research Quarterly,* vol. 41, 1970, pp. 615-18.

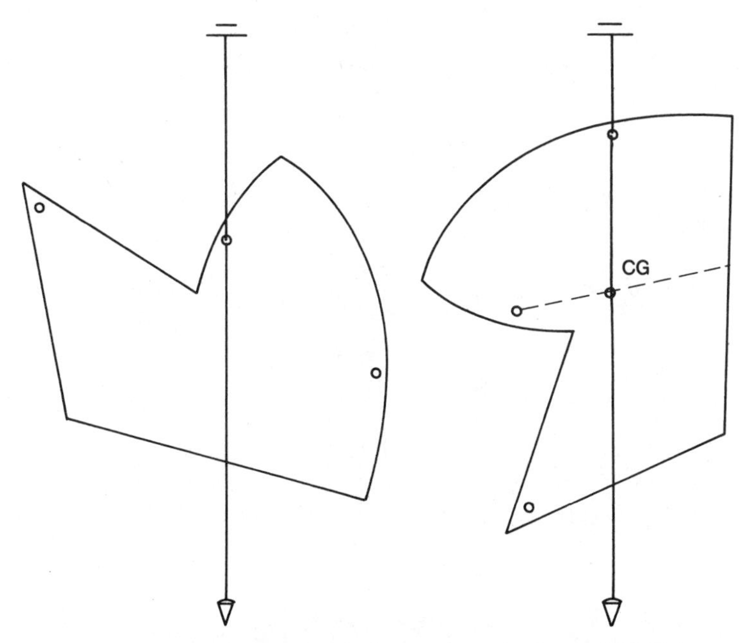

Figure 1-6 Location of the center of gravity for a rigid, flat board. The board is suspended from at least two points, and a plumb line is marked on the board. The place of intersection of the drawn lines is the board's center of gravity or balance point.

ing segmental data, and others have worked with reaction boards to obtain mass centers of bodies in different positions. Interested students can find good accounts of the various methods of obtaining center of gravity in Hay,[5] LeVeau,[6] and Plagenhoef.[7] Research continues, but for the present we have to be content with using figures which represent an average human structure as determined to date.

For the average human in an erect stance, the center of mass is typically found in the range of 55 to 60 percent of standing height, or approximately on a level with the second sacral vertebra (Figure 1-7). It is higher in children, and slightly lower in women than in men. In addition, each segment of the body has its own mass center, which must be reckoned with in most types of movement analysis. There are fourteen major body segments, sometimes referred to by engineers as *links,* to consider: head, trunk, and two each of thighs, forelegs, feet, upper arms, forearms, and hands. Toes and

[5] James Hay, *The Biomechanics of Sports Techniques,* 2nd ed. (Englewood Cliffs, N.J.: Prentice-Hall, Inc., 1978), pp. 127-38.

[6] Barney LeVeau, *Williams and Lissner: Biomechanics of Human Motion* (Philadelphia: W. B. Saunders Company, 1977), pp. 79-88.

[7] Stanley Plagenhoef, *Patterns of Human Motion* (Englewood Cliffs, N.J.: Prentice-Hall, Inc., 1971), pp. 21-27.

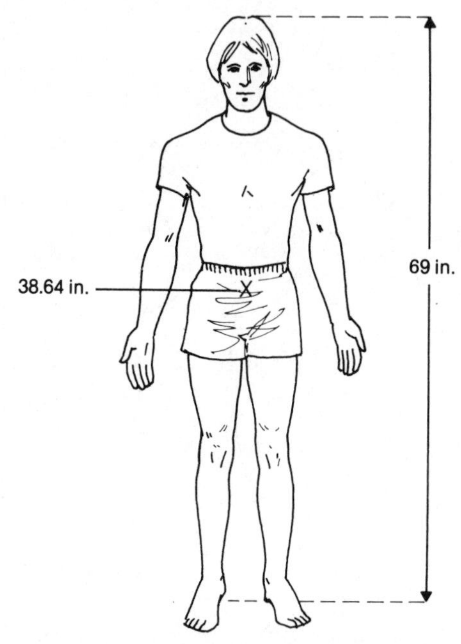

Figure 1-7 The approximate location of the center of gravity of an average male is at a level which is 56 percent of his standing height.

fingers are normally insignificant factors in motion analysis and are not considered separately. Roebuck provides a table which compiles segmental mass-body mass ratios from a number of cadaver studies dating from 1860.[8]

In describing segment movement, it is important to note whether segment *A* is moving while an adjoining segment *B* is stationary, whether *B* is moving while *A* is stabilized, or whether both *A* and *B* are in motion. For example, there can be flexion at the hip as the trunk moves down while the thigh is stationary, as in a toe touch. Or, as in a high kick, the flexion can result as the thigh moves toward the trunk. Finally, both segments may be in motion, as in a stooping knee bend or in a pike position springboard dive. Similarly, in the pullup exercise, the humerus moves toward a stable forearm, while in barbell curls, the forearm moves but the humerus is relatively still.

To help illustrate how a human differs from a solid mass and why segmental understanding is important, note in Figure 1-8*a* that unless some force acts on it, the solid mass will remain on the table indefinitely, since its mass center is not beyond the table's edge. In Figure 1-8*b*, observe that a human can likewise remain in a fixed position on the table if the necessary

[8] J. A. Roebuck, Jr., K. H. E. Kroemer, and W. G. Thomson, *Engineering Anthropometry Methods* (New York: John Wiley & Sons, Inc., 1975), p. 58.

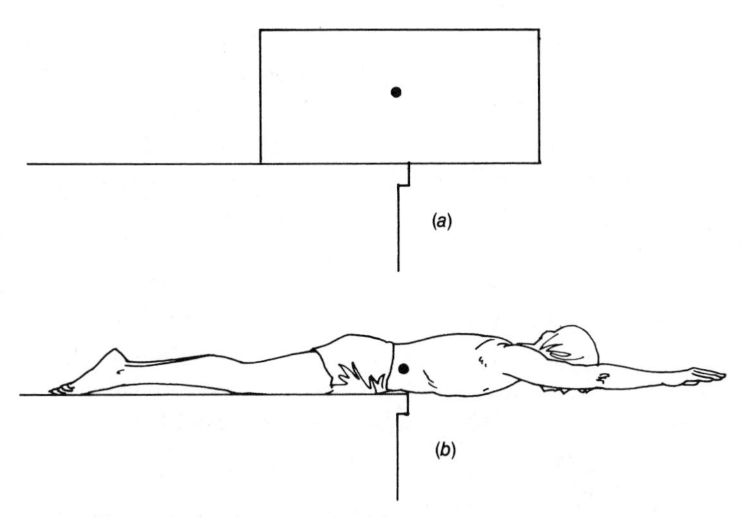

Figure 1-8 Comparison of the effects of gravity on a solid mass (*a*) and a human body (*b*) supported by a table.

muscles are statically contracted to hold this position. However, as the muscles will inevitably tire, those parts of the body that are not supported by the table will sag and drop, moving the mass center toward the feet while the lower body remains on the table. Gravity acts identically in both cases, but the difference, of course, is that the human has joints, where the downward pull of gravity must be actively resisted by muscular forces, whereas the solid mass is passively held together by some cohesive qualities of the material.

Knowledge of the exact location of the center of gravity in a performing athlete is not vital for the teacher, coach, or athlete. First of all, it is very difficult to determine at any instant, and second, its location may very well be shifting throughout the performance of the skill. An understanding of the *role* played by center of gravity in sports *is* important, and a number of principles which will be presented require knowledge of center of gravity.

SUGGESTED READINGS

ALLEY, LOUIS E., "Utilization of Mechanics in Physical Education and Athletics," *Journal of Health, Physical Education, and Recreation,* vol. 37, March 1966, pp. 67–70.

ASMUSSEN, E., and K. JORGENSEN, eds., *Biomechanics VI,* 2 vols. (Baltimore: University Park Press, 1978).

CERQUILIGLINI, S., A. VENERANDO, and J. WARTENWEILER, eds., *Biomechanics III* (Basel, Switzerland: S. Karger AG, 1973).

CLARYS, J., and L. LEWILLIE, eds., *Swimming II,* International Series on Sciences (Baltimore: University Park Press, 1975).

CLAUSER, C. E., J. T. MCCONVILLE, and J. W. YOUNG, "Weight, Volume, and Center of Mass of Segments of the Human Body," AMRL-TR-69-70 (Wright-Patterson Air Force Base, Ohio: August 1969).

COOPER, J. M., ed., *Selected Topics on Biomechanics: Proceedings of the C.I.C. Symposium on Biomechanics* (Chicago: Athletic Institute, 1971).

HARRIS, RUTH W., *Kinesiology: Workbook and Laboratory Manual* (Boston: Houghton Mifflin Company, 1977).

KELLEY, D. L., "Supporting Biomechanics Subject Matter in the Undergraduate Curriculum," in *Kinesiology IV* (Washington: American Association for Health, Physical Education, and Recreation, 1974).

KOMI, P. V., ed., *Biomechanics V-A* and *V-B* (Baltimore: University Park Press, 1976).

LOCKE, L. F., "Kinesiology and the Profession," *Journal of Health, Physical Education, and Recreation,* vol. 36, September 1965, p. 69.

NELSON, R. C., and C. A. MOREHOUSE, eds., *Biomechanics IV* (Baltimore: University Park Press, 1974).

VREDENBREGT, J., and J. WARTENWEILER, eds., *Biomechanics II* (Basel, Switzerland: S. Karger AG, 1971).

WARTENWEILER, J., E. JOKL, and M. HEBBELINCK, eds., *Biomechanics: Technique of Drawings of Movement and Movement Analysis* (Basel, Switzerland: S. Karger AG, 1968).

CHAPTER 2

Basic mathematics

An understanding of the underlying mathematics will aid the student in fully grasping a number of the principles in biomechanics. This chapter covers measurement systems, mass, weight, vectors, scalars, fundamental trigonometry, and triangle laws.

MEASUREMENT SYSTEMS

Although the universally accepted metric system (Système International d'Unités, or SI system) is expected to be adopted for common use in the United States at some future date, this text will continue to employ the more familiar British units of measurement, because most of this nation's playing fields, tracks, courts, and pools are laid out in feet and yards, weight classes in boxing and wrestling are in pounds, and athletes' heights are recorded in feet and inches. Metric equivalents will be noted frequently throughout the text for the benefit of those readers who prefer to use them, and some of the worked examples and end-of-chapter problems will be in SI units.

In the British system,

Force is expressed in pounds (lb).

Time is noted in seconds (s), minutes (min), or hours (h).

Length is measured in feet (ft), yards (yd), or miles (mi).

Mass is indicated in slugs.

In the MKS metric system (meter-kilogram-second),

Force is given in newtons (N).
Time is in seconds (s).
Length is measured in meters (m).
Mass is expressed in kilograms (kg).

In the less-often used CGS system (centimeter-gram-second),

Force is measured in dynes (dyn).
Time is indicated in seconds (s).
Length is in centimeters (cm).
Mass is expressed in grams (g).

It is most important to stay wholly within one system of measurements in making computations. While the student may be familiar with all of the measurement systems, care must be taken not to mix the units.

WEIGHT AND MASS

In physics, weight and mass are two distinctly different terms, and they cannot be used interchangeably. It is necessary to be aware of the differences, particularly in using certain formulas.

Mass is a numerical measure of an object's inertia, and it is sometimes described as the quantity of matter in a body. Mass, unlike weight, does not vary from place to place on earth. Weight is a downwardly directed force and may be defined as the gravitational attraction of the earth for a given object.

Confusion arises when we must distinguish between an object's weight and its inertia, the first being expressed in newtons and the latter in kilograms. A 1-kg mass "weighs" 9.8 N, but as Wolfe states, "When you shop for hamburger at the local market, you want to buy a kilogram of meat rather than 9.8 newtons of meat, even if you are a professor of physics, and you expect the butcher to 'weigh' it for you."[1]

Weight is commonly measured by means of a spring scale calibrated in pound and ounce units. It is found as the product of mass and gravity; thus

[1] Hugh C. Wolfe, "The Weight-Mass Controversy," *American Journal of Physics,* vol. 47, no. 7, July 1979, p. 574.

$$W = mg \tag{2-1}$$

where m = mass (expressed in the British system in slugs and in the SI system in kilograms)

g = the gravitational constant (32 ft/s² in the British system and 9.8 m/s² in the SI system)

W = weight (expressed in pounds in the British system and in newtons in the SI system)

To determine mass in the British system, Eq. (2–1) is rearranged to divide an object's weight in pounds by 32 ft/s², and the resulting derived unit is the *slug*. One slug is defined as that mass which experiences an acceleration of one foot per second per second when acted upon by an external unbalanced force of one pound. Thus, a 1-slug mass may be said to equal a 32-lb weight. The slug is a more convenient unit to use than the one which evolves from the division of weight by gravity as:

$$\frac{W}{g} = \frac{\text{lb}}{\text{ft/s}^2} = \frac{\text{lb} \cdot \text{s}}{\text{ft/s}} = \frac{\text{lb} \cdot \text{s}^2}{\text{ft}}$$

In deriving slugs, it is important to use pounds rather than ounces. Divide the number of ounces by 16 to obtain the number of pounds.

Example 2–1. What is the mass of a 160-lb man?

$$m = \frac{W}{g} = \frac{160 \text{ lb}}{32 \text{ ft/s}^2} = 5 \text{ slugs}$$

In SI units, 1 lb = 0.4536 kg; thus

$$160 \times 0.4536 = 72.6 \text{ kg}$$

Example 2–2. What is the mass of a 6-oz ball?

Convert to pounds by

$$\frac{6 \text{ oz}}{16 \text{ oz/lb}} = 0.375 \text{ lb}$$

$$m = \frac{W}{g} = \frac{0.375 \text{ lb}}{32 \text{ ft/s}^2} = 0.0117 \text{ slugs}$$

In the MKS system, one kilogram is that mass which experiences an acceleration of one meter per second per second when acted upon by an external force of one newton.

VECTORS AND SCALARS

Any physical quantity, such as velocity, acceleration, force, or displacement, which has both magnitude and direction is termed a *vector quantity*. For example, it is not sufficient to be told that the magnitude of the net force applied to an object is 24 lb. If that force were applied vertically downward to the object, the object would remain stationary, whereas a horizontal application would accelerate the object in the direction of application. It is necessary to know in which direction a force is being applied in order to judge what effects the force will have on the type and direction of the resulting movement. Similarly, speed alone may not be sufficiently descriptive, and it may be necessary to specify the direction of the motion, in which case the vector term *velocity* is used. In the case of projectiles, direction also requires a specification of the angle of release.

Magnitudes consist of a number and a unit, for example, 7 lb, 3 ft, or 20 ft/s. Direction may be indicated in the usual compass terms or in degrees from some reference line.

Those quantities which are fully described by their magnitudes alone are called *scalars*. In scalar measurement, direction is of no consequence. Examples of scalar quantities are speed (25 mi/h), distance (2 mi), time (3 h), mass (6 slugs), and temperature (98.6°). Scalars can be arithmetically added, subtracted, divided, and multiplied, whereas vectors must be treated geometrically with triangles and parallelograms, except when they happen to fall on the same line, as do the forces depicted in Figure 2-1. Forces applied in a tug of war are all acting in the same line, some positively to the right and some negatively to the left. The winning team will have exerted a greater net force, which includes not only the combined weight and strength of the members but also the friction between their feet and the ground and between their hands and the rope.

Figure 2-1 Arithmetic addition of force vectors which fall on the same line. Subtracting 4 lb in the negative direction from the 12 lb total force acting in the positive direction leaves an unbalanced force of 8 lb in the positive direction.

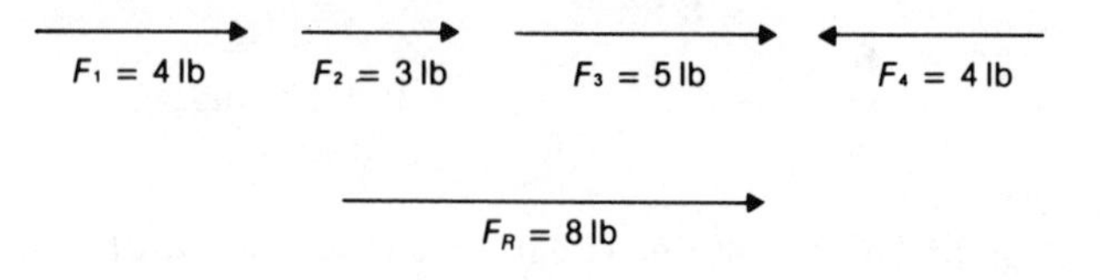

For mathematical purposes, a vector is a directed straight line, an arrow which must always be drawn to a precise scale and at exact relative angles to accurately represent the magnitude (by its length) and the direction (as indicated by the arrowhead).

The sum of two or more vectors is known as a *resultant,* which is simply a single vector that has the same effect as would all of the component vectors acting together. The length of the resultant vector describes its magnitude in proportion to the magnitudes of the components, and of course its direction is evident from its angle.

When two component vectors are known, a resultant vector can be found by the method known as a *composition* of vectors. When the resultant is given and the two components are being sought, a process called *resolution* of vectors is used.

In the composition of vectors, the order of placement of the arrows makes no difference so long as they are placed head to tail (Figure 2–2*a*). But note that in the figure, A plus B or A plus B plus C does not simply add up to R. The expression *polygon of vectors* may be used to describe the diagraming of more than two vectors to be added, as in Figure 2–2*b*. Furthermore, the vectors being added do not have to be in the same plane, but for the sake of simplicity two-dimensional motion will be assumed in examples.

When two component vectors and the angle between them are known, a *parallelogram* can be constructed by placing the vectors tail to tail and then drawing lines parallel to the vectors from the barb of each arrow. A diagonal line drawn from the common origin of the components to the opposite corner represents the resultant magnitude and direction, and it should be evident that such a parallelogram is really nothing more than the addition or composition of two vectors laid head to tail (Figure 2–3).

Figure 2–2 (*a*) Composition of two vectors (*b*) Composition of three vectors

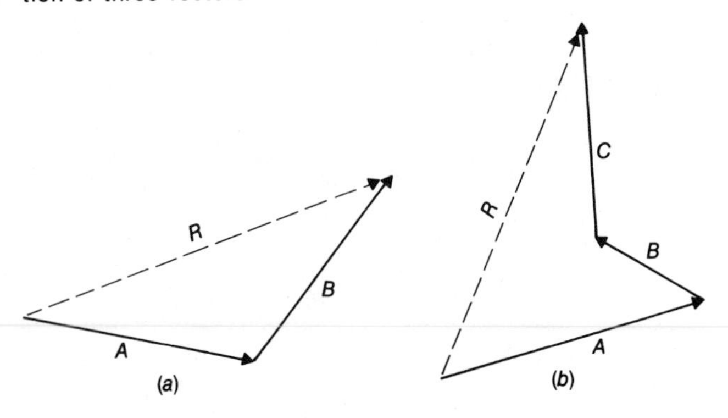

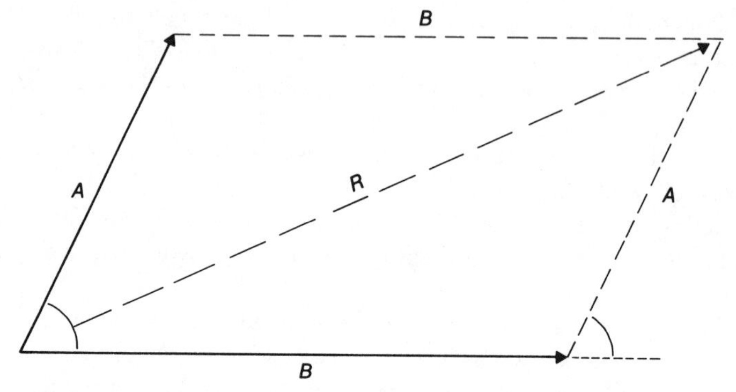

Figure 2-3 Composition of two vectors by the parallelogram method

Vector Problems

While many of the following example problems may not have wide applicability to sports analysis, they are included here to give the student practice in handling vector problems. Some knowledge of elementary trigonometry is presumed, but those who need review should refer to Appendix B before proceeding further.

Displacement, or change of position in space, is a vector quantity involving, as it does, a distance measure and a direction. A woman might take a 14-mi hike in either a straight line or a path involving one or more changes of direction. If the interest is only in how many miles she has walked, then of course the distance is 14 mi, a scalar quantity. However, if the concern is with the hiker's displacement, that is, how far and in what direction she is relative to her starting point, then we are dealing with a vector quantity and must plot the trip.

Suppose that, as in Figure 2-4, the hiker walks 6 mi east and then makes a 90° turn to the north for the remaining 8 mi. Since this is now conveniently a right-triangle problem, the hiker's displacement vector must be determined by using the Pythagorean theorem, which states that in any right triangle the square of the hypotenuse is equal to the sum of the squares of the two sides, or

$$c^2 = a^2 + b^2 \tag{2-2}$$

Thus, in Figure 2-4,

$$\begin{aligned} \text{Displacement}^2 &= (6\ \text{mi})^2 + (8\ \text{mi})^2 \\ \text{Displacement} &= \sqrt{(36 + 64)\ \text{mi}^2} \\ &= 10\ \text{mi northeast} \end{aligned}$$

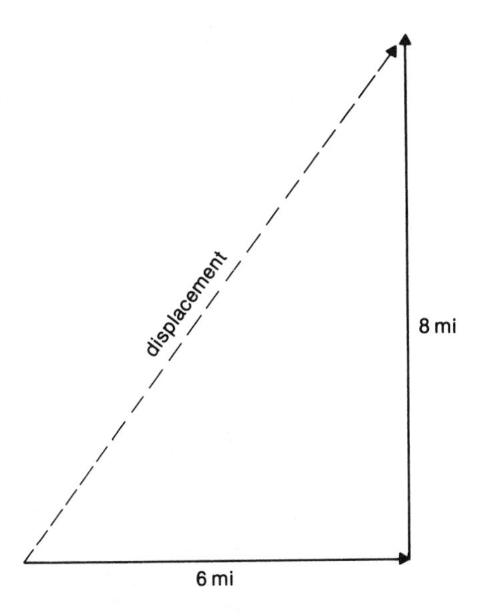

Figure 2-4 A right triangle vector problem in which two sides are known and a resultant hypotenuse is sought.

Note that the answer includes both a magnitude and a direction, and it should be clear that this magnitude cannot be obtained by simply adding two vectors having lengths of 6 mi and 8 mi.

This problem can also be solved by a graphic method, which requires the careful laying out of the vectors to exact scale on graph paper. The displacement resultant and its angle can then be measured directly from the drawing with a ruler and a protractor (Figure 2-5).

Now what if we are faced with a displacement problem which does not involve an apparent right triangle? For example, suppose a man on a bicycle travels east for 12 mi and then heads north of east at a 53° angle for an additional 25 mi. What is his displacement? We could proceed graphically, as in Figure 2-5, and take our measurements directly from the triangle as drawn. Or we could create right triangles *ADC* and *BDC* as shown in Figure 2-6 and find the resultant vector *AC* trigonometrically.

The following information has been given: side $AB = 12$ mi, side $BC = 25$ mi, and angle *B* is 53°. In the right triangle *BDC* we have created, the hypotenuse is known to be 25 mi and the angle *B* to be 53°, so it is necessary to find the lengths of sides *BD* and *CD*, the adjacent and opposite sides, respectively. *BD*, the adjacent side, is obtained by the function definition:

$$\text{cosine} = \frac{\text{adjacent side}}{\text{hypotenuse}}$$

$$\cos 53^\circ = \frac{BD}{25\text{ mi}}$$

$$.6018 = \frac{BD}{25\text{ mi}}$$

$$BD = 15\text{ mi east}$$

CD, the opposite side, is obtained by the definition:

$$\text{sine} = \frac{\text{opposite side}}{\text{hypotenuse}}$$

$$\sin 53^\circ = \frac{CD}{25\text{ mi}}$$

$$.7986 = \frac{CD}{25\text{ mi}}$$

$$CD = 20\text{ mi north}$$

Figure 2-5 Graphic method for the composition of vectors

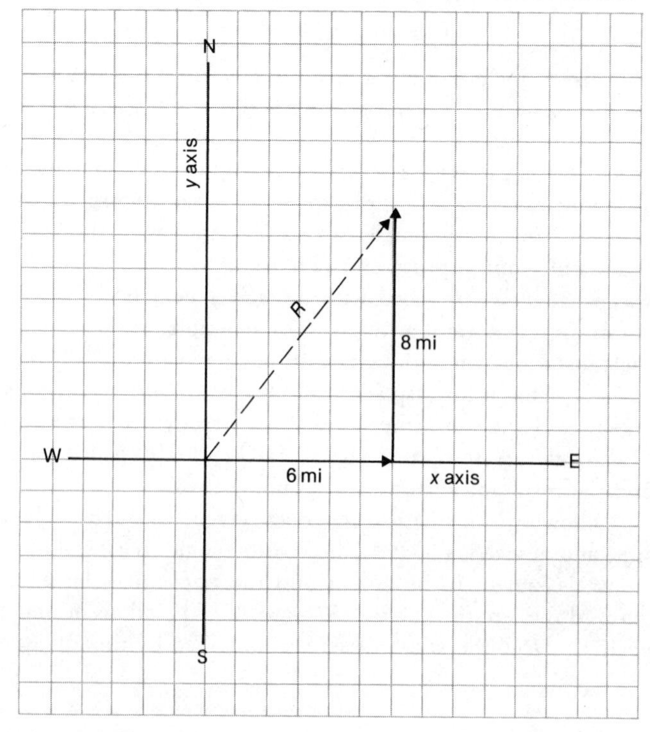

Now in the larger created right triangle *ADC*, the lengths of *AB* and *BD* are added together to find side *AD* (12 + 15 = 27), and side *CD* was just found to be 20 mi. With this information, the Pythagorean theorem is used to determine the displacement resultant side *AC*:

$$\begin{aligned} AC^2 &= AD^2 + CD^2 \\ &= (27 \text{ mi})^2 + (20 \text{ mi})^2 \\ &= (729 + 400)\text{mi}^2 \\ AC &= \sqrt{1129 \text{ mi}^2} \\ &= 33.6 \text{ mi northeast} \end{aligned}$$

Angle *A* formed between the displacement vector and the vector from the origin may be found by the function definition:

$$\begin{aligned} \tan\theta &= \frac{\text{opposite side}}{\text{adjacent side}} \\ &= \frac{20 \text{ mi}}{27 \text{ mi}} = 0.74 \end{aligned}$$

which is approximately the tangent of 37° (note that the Greek letter θ [theta] will be used for the angle sought).

Thus far, triangle problems have been solved by utilizing trigonometric and graphic methods. There is also a mathematical approach in which it is unnecessary to actually draw vectors. Two laws are of particular use.

The Law of Cosines

The *cosine law* may be used to solve any triangle problem in which two sides and an included angle are known and the third side is sought. It states:

> *The square of any side of a triangle is equal to the sum of the squares of the other two sides minus twice the product of those two sides and the cosine of the angle included by them.*

This definition is represented by the equation

$$c^2 = a^2 + b^2 - 2ab\cos\theta \qquad \textbf{(2–3)}$$

where c is the unknown side and θ is the angle opposite the unknown side included between sides a and b when these sides are arranged head to tail.

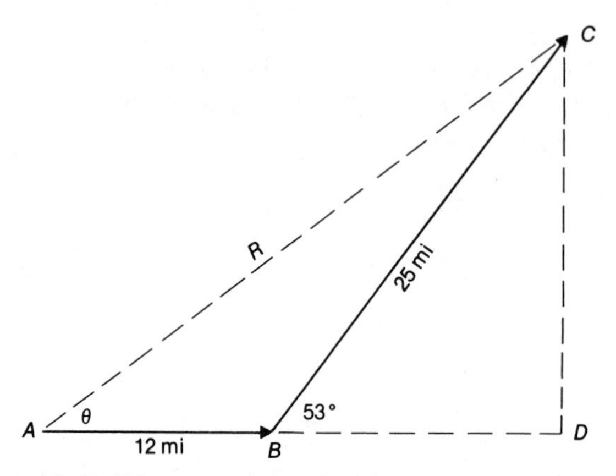

Figure 2-6 Trigonometric method for determining the resultant of an oblique triangle.

Example 2–3. Given the data in Figure 2–6, solve for AC using the cosine law.

Side AB is 12 mi, side BC is 25 mi, the included angle ABC is 127° (which was found by subtracting 53° from 180°). These figures now can be substituted into Eq. (2–3):

$$AC^2 = (12 \text{ mi})^2 + (25 \text{ mi})^2 - (2)(12 \text{ mi})(25 \text{ mi}) \cos 127^\circ$$

Still lacking is the cosine of 127°, and this is found by subtracting 127° from 180°. Appendix E indicates that the cosine of an obtuse angle, one larger than 90°, equals the cosine of its supplement but is opposite in sign. The supplement of 127° is 53°, and the cosine of 127° is negative.

$$\cos 127^\circ = -\cos(180^\circ - 127^\circ) = -\cos 53^\circ = -.6018$$

$$AC^2 = [144 + 625 - (600)(-.6018)]\text{mi}^2$$

$$= 769 \text{ mi} - (-361 \text{ mi})$$

In subtracting a negative number, change the sign and add.

$$AC^2 = 1130 \text{ mi}^2$$

$$AC = \sqrt{1130 \text{ mi}^2} = 33.6 \text{ mi northeast}$$

If the vectors are depicted tail to tail, or if one is accustomed to using the supplement automatically whenever an obtuse angle is involved, then the cosine law can be written

$$c^2 = a^2 + b^2 + 2ab(\cos\theta) \tag{2-4}$$

The cosine law can also be used to find any angle of an oblique triangle when all three sides are known.

$$\cos C = \frac{a^2 + b^2 - c^2}{2ab} \tag{2-5}$$

This is simply a rearrangement of Eq. (2-3) to solve for angle C.

The Law of Sines

The *sine law* can be used to find an angle when two sides and an opposite angle of a triangle are known. When one side and two angles are known, the sine law can be used to find another side. The sine law states:

In any triangle, the sides are proportional to the sines of the opposite angles.

This definition is represented by the equation

$$\frac{\text{Side } a}{\sin A} = \frac{\text{Side } b}{\sin B} = \frac{\text{Side } c}{\sin C} \tag{2-6}$$

Referring again to Appendix E, note that the sine of any angle up to 180° is positive in sign or, restating, the sine of an obtuse angle equals the sine of its supplement.

Example 2-4. In Fig. 2-6, having found resultant AC to be 33.6 mi and knowing BC to be 25 mi, find angle A, which is opposite side BC.

Use Eq. (2-6):

$$\frac{BC}{\sin A} = \frac{AC}{\sin B}$$

$$\frac{25 \text{ mi}}{\sin A} = \frac{33.6 \text{ mi}}{\sin 127^\circ}$$

$$\frac{25}{\sin A} = \frac{33.6}{.7986}$$

(continued)

$$33.6 \sin A = 25 \times .7986$$

$$\sin A = \frac{19.97}{33.6} = .594$$

which is approximately the sine of 37°.

UNIT CONVERSIONS

It is often necessary to convert from one system of measurements or one set of units to another. The longer of the two methods to be used is the unity method, which changes the units but not the values. To obtain feet per second when miles per hour are given, multiply the given measure, say 60 mi/h, by a distance ratio and a time ratio which are in the desired units. Then by canceling out the undesired units, the conversion to feet per second is completed.

Example 2–5. Convert 60 mi/h to its equivalent in feet per second.

$$\frac{60 \text{ mi}}{1 \text{ h}} \times \frac{5280 \text{ ft}}{1 \text{ mi}} \times \frac{1 \text{ h}}{3600 \text{ s}} = \frac{60 \times 5280 \text{ ft}}{3600 \text{ s}}$$
$$= 88 \text{ ft/s}$$

Knowing that 60 mi/h equals 88 ft/s is handy because of its easy divisibility and use in quick estimates of velocity equivalents. One immediately sees that 30 mi/h equals 44 ft/s, 15 mi/h equals 22 ft/s, and so on.

Example 2–6. Convert 17 ft/s into miles per hour.

$$\frac{17 \text{ ft}}{1 \text{ s}} \times \frac{1 \text{ mi}}{5280 \text{ ft}} \times \frac{3600 \text{ s}}{1 \text{ h}} = \frac{61{,}200 \text{ mi}}{5280 \text{ h}}$$
$$= 11.6 \text{ mi/h}$$

Example 2–7. Convert 30 m/s into feet per second.

$$\frac{30 \text{ m}}{1 \text{ s}} \times \frac{3.28 \text{ ft}}{1 \text{ m}} = 98.4 \text{ ft/s}$$

Since in the unity method one must already know or look up comparable unit values, it would be a shorter process to find conversion factors in some table such as Table 2–1. A more complete list of unit relations may be found in Appendix D.

TABLE 2-1
Some Common Distance Unit Conversion Factors

Given	*Desired*	*Multiplier*
mi/h	ft/s	1.467
ft/s	mi/h	0.68
m/s	ft/s	3.28
ft/s	m/s	0.305
km	mi	0.6214
mi	km	1.609

Example 2-8. Convert 26 mi into kilometers.

$$26 \text{ mi} \times 1.609 \text{ km/mi} = 41.83 \text{ km}$$

Example 2-9. Convert 11 mi/h into feet per second.

$$11 \text{ mi/h} \times \frac{1.467 \text{ ft/s}}{\text{mi/h}} = 16.14 \text{ ft/s}$$

Example 2-10. Convert 100 m into yards.

$$100 \text{ m} \times 3.28 \text{ ft/m} \div 3 \text{ ft/yd} = 109.33 \text{ yd}$$

SUMMARY AND DISCUSSION

The meter-kilogram-second system (MKS) of measurement is the SI international system used almost everywhere in the world. Its adoption in the United States is proceeding slowly and has met considerable resistance. For this reason, the more familiar British units have been emphasized in this text.

An understanding of the differences between vector and scalar quantities is important to the student. Quantities such as force and velocity are best described in terms of both magnitude and direction. Scalar quantities such as speed and height need be expressed only in terms of a magnitude. The mathematical treatment of vectors is different from that needed for scalars, and a working knowledge of at least basic trigonometry is needed when dealing with vectors.

Two useful laws are the sine law, Eq. (2-6), and the cosine law, Eq. (2-3), which are applied in non-right-triangle problems. The sine law may be used to find an unknown angle when any two sides of a triangle and one angle opposite either of these sides are known. Or an unknown side may be found if two sides and another angle are known. The cosine law requires that two sides and an *included* angle be known in order to find the third side.

It was suggested earlier that a student who has not had some trigonometry should study the material in Appendix B before undertaking the right-triangle problems. However, the philosophy of this text is that any use of mathematics is intended to simply reinforce many of the principles and laws of mechanics as they seem to apply to sports. Some familiarity with basic mathematics is necessary if the student is to grasp the meaning of the vast amount of biomechanical research that is being published in the technical literature and being interpreted in the popular literature.

Problems for the Student

1. Express 50 mi/h in feet per second.
2. Compute the mass of a 200-lb man.
3. What is the displacement of a boat which leaves its dock heading west for 7 mi and then changes course 37° to the north for 5 mi more? Solve trigonometrically.
4. What is the weight of a 2.5-slug mass?
5. Using the sine law, find angle B in an oblique triangle ABC where side $a = 13$, side $b = 20$, and angle $A = 20°$.
6. For the triangle in problem 5, find side c using both the sine law and the cosine law.
7. For a right triangle ACB, find side b if side $a = 21$ and the hypotenuse $c = 35$.

(Answers to odd-numbered problems may be found in Appendix F.)

CHAPTER 3

Force

What is a force? You cannot *see* a force that is acting on some object, but you can certainly feel a force if it acts on you. The only evidence we generally have that a force is acting is by the sound made upon impact or by the fact that an object is moved or distorted.

Force is most simply defined as a directed push or pull by one body acting upon another. Because it has both magnitude and direction, it is classified as a vector quantity. All changes in motion are due to some force action, but not all force actions result in changes in motion of the body acted upon. It is necessary that there be an *unbalanced* force, that is, one whose magnitude is greater than the magnitude of the object to be moved. The study of the effects of forces has previously been labeled *kinetics.*

A golf ball is put into flight by a force applied to it by a swung club, which in turn has received its motion from various muscle forces acting to move the golfer's trunk, shoulders, arms, and wrists. As soon as the ball leaves the tee, it encounters air resistance, which begins to reduce its velocity, while at the same time the force of gravity immediately acts to limit the ball's ascent and to cause it to fall back to earth.

In addition to being able to push or pull, an unbalanced force may also cause distortion in an object. When the force tends to stretch an object, it is labeled a *tension* force. When its effect is to squeeze an object, it is known as a *compression* force. A *shear* force is one which causes one part of an object to slide relative to another part.

UNITS OF FORCE

In the British system of measurement, the unit for force is the familiar pound, while in the SI, or meter-kilogram-second, system it is the newton. In the infrequently used CGS, or centimeter-gram-second, system, force is measured in dynes.

A *pound* (lb) is that amount of force which will accelerate a mass of one slug one foot per second per second.

$$1 \text{ lb} = 1 \text{ slug} \cdot \text{ft/s}^2$$

A *newton* (N) is that amount of force which will accelerate a mass of one kilogram one meter per second per second.

$$1 \text{ N} = 1 \text{ kg} \cdot \text{m/s}^2$$

A *dyne* (dyn) is that amount of force which will accelerate a mass of one gram one centimeter per second per second.

$$1 \text{ dyn} = 1 \text{ g} \cdot \text{cm/s}^2$$

EFFECTS OF FORCE

Whether or not a desired motion results from the application of a force depends upon several considerations:

1. A change in motion will result only if the magnitude of the force exceeds the magnitude of the resistance afforded by the body; that is, an unbalanced force must be applied. If the body resists motion with a force of 8 lb and the applied force is 10 lb, an unbalanced or net force of 2 lb remains and motion will result. The most common resistances are friction, a force that always opposes motion of one body over another, and inertia, which is a body's tendency to resist any change in its motion and which is proportional to the weight or mass of the body.

2. The direction in which a force is being applied to a body will determine whether the resulting motion is linear, angular, or a combination of these. All the forces depicted as vectors in Figure 3-1 are identical in magnitude but different in direction, so that when each is applied separately at the same point on the body, each will produce a different kind of motion. The effect of force F_1 acting alone will be to move the body in a linear direction, as represented by the dotted-line extension of the F_1 vector, and at the same time to cause the body to rotate in a clockwise manner, because the ex-

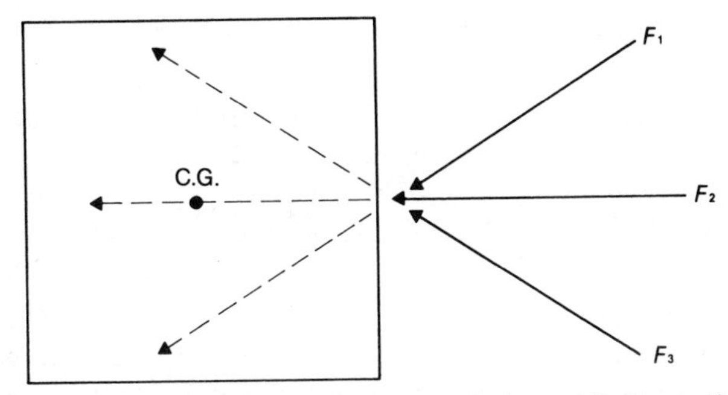

Figure 3–1 Application of identical unbalanced forces to the same point on a body. Each applied separately produces a different motion result.

tended action line of the vector passes below the center of gravity of the body. (Any such force that does not pass through an object's rotational axis is termed an *eccentric force.*) Force F_3 is also eccentric, and because its action line passes above the center of gravity, which in these examples is the axis for rotation, it will cause counterclockwise rotation along with linear motion in the direction of force application. The action line of force F_2 passes through the object's mass center and therefore will produce only linear motion to the left.

3. The point of application of a force is another determinant of whether the body will move in a linear or an angular manner. All the force vectors shown in Figure 3–2 are identical in magnitude and direction, but they are applied at different points on the body. The eccentric, unbalanced force F_1 will cause both counterclockwise and linear motion. Being also eccentric, force F_3 will also make the body move in a linear manner, while at the same time causing it to rotate clockwise. Since force F_2 passes through the

Figure 3–2 Application of three identical unbalanced forces to three different points on a body. As each is applied separately, a different motion results.

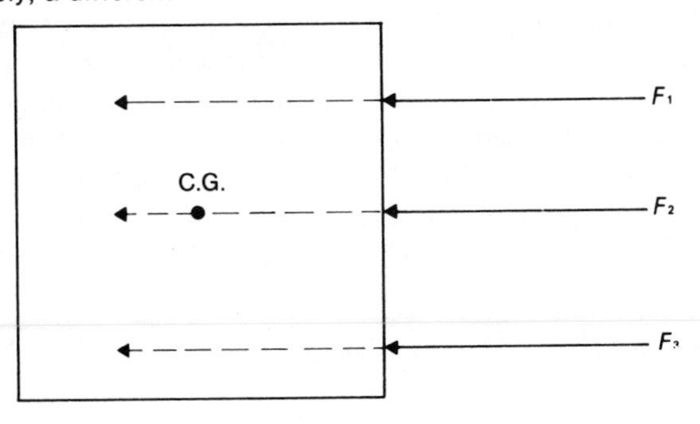

axis, the body's mass center, the body will be moved to the left in a linear path.

4. The extent of any change in velocity experienced by a body is proportional to the magnitude of the unbalanced force, as already mentioned, and also to the length of time or to the distance over which the force is applied. The longer a force can be applied, the more a body will accelerate. This principle will be explained in Chapter 4 in the section on Newton's laws of motion.

5. If several forces are to be successively applied in the same direction to a body, the sequence in which these forces are applied will affect the resulting movement. As Bunn points out, if the intent is to give an object maximal velocity, each new force should be timed to be applied at the point where the preceding force has made its greatest contribution toward increasing the object's velocity.[1] This point will coincide with that of the reduction of the acceleration attributable to the preceding force. While the principle stated in this way may be overly simplified, it generally holds that in a variety of power-type activities—such as shot-putting, discus throwing, and batting—the larger and slower-contracting muscles will initiate movement and, at some optimal moment, the weaker but faster muscles will be brought successively into play. As the role of each muscle group toward acceleration is completed, the muscles continue to apply some force to provide an effective reaction to the new forces brought into action.

Of the principles just listed, number 5, regarding the summation of forces, is the most complex. The fine timing of various movements is the basis for good sports technique, and it is extremely difficult for a coach to visually detect at which point in a skill an athlete is improperly applying a force. Motion pictures or videotapes can be of some help in such analyses. For a researcher, an analysis of a power event must take into account, or should at least recognize, that the force capabilities of a muscle are reduced as its contraction accelerates, and that there are reactions between adjacent moving segments of varying masses.

All five principles of force application effectiveness are well illustrated in the shot put. The *direction* in which force is applied by an athlete to the shot determines the angle of its release into the air, as well as its velocity. The final *point of application* of the force by the wrist and fingers establishes whether or not the shot will rotate in flight and in which direction it will rotate. The longer the *time* in which the putter can apply an unbalanced force to the shot, the greater the final velocity of the shot. Distance over which force is applied is related to the time. All of the putter's forces should be *summated*

[1] John Bunn, *Scientific Principles of Coaching* (Englewood Cliffs, N.J.: Prentice-Hall, Inc., 1972), p. 77.

from the ground up; that is, the larger muscles of the legs and then of the trunk initiate the motion across the circle and get the necessary reaction from the ground and from various body segments. Additional muscular forces are applied successively to the shot until it is released with a final flick of the wrist and fingers.

As will be elaborated in the next chapter, any acceleration a body experiences is proportional to the force that is applied to it. This is equally true for negative acceleration, because when a moving object is slowed down, some braking force is required and its magnitude determines the deceleration rate.

SOURCES OF FORCE

In human movement, contracting muscles provide an *internal* force system as they pull on the various bony levers. A muscle pulls equally hard on its origin on one bone and its insertion on another. One segment is generally stabilized by its own mass or by the action of other fixator muscles, while the other segment being acted upon is free to move. When a muscle shortens to cause movement, it is called an *agonist,* or mover, and it is said to undergo a *concentric* contraction. In situations where the muscle maintains tension while lengthening against a resistance such as gravity, the contraction is termed *eccentric* (not to be confused with *eccentric force,* which produces rotation).

Eccentric contractions serve to control the descent of a body segment or an external weight so that the movement is slower than it would be if it were falling at the gravitational acceleration of 32 ft/s^2 or 9.8 m/s^2. This is certainly necessary in an exercise such as the pullup, wherein the elbow-flexing muscles, which have raised the body by concentrically contracting (shortening), now lower the body slowly by eccentrically contracting to allow the elbows to become extended. Without a controlled eccentric contraction, the body would fall to the extended-arm position and the resulting jerk could injure the joints and their supporting tissue.

Concentric and eccentric contractions involve changes in muscle length and for that reason can be classified as *isotonic* contractions. Most exercises and sports skills utilize isotonic contractions of muscles to effect movement. In an *isometric* contraction, the muscles exert maximum tension but no motion results, because the magnitude of the resistance is too great. Such contractions may occur when one attempts to lift a barbell that is too heavy or when one wrestler tries mightily to pin an opponent who resists with an equal force. Another type of contraction that produces no motion is termed a *static* contraction. Here a person exerts only enough tension to hold a body segment in a particular position for a period of time, as when a fencer holds the rear arm with the elbow at shoulder height.

In making anatomic analyses, it is important to be able to identify the muscles that are acting, if not by name then at least by broad group, such as hip flexors and elbow extensors. This identification is not possible if it is not first established whether gravity is playing a role in the movement. For example, in flexing the leg at the hip, the hip flexor muscles are the agonists while the hip extensors are antagonists. In a slow lowering of the leg against the pull of gravity, the hip flexors now eccentrically contract and the extensors are again inactive. However, if from the flexed position you wish to bring the leg back very quickly, then the hip extensor group is acting agonistically while the flexor group acts antagonistically by relaxing. But in medial and lateral rotational movements of the leg, gravity plays no role, and so the medial rotators and outward rotators each in turn act as agonists and concentrically contract to medially and outwardly rotate the leg.

There are a number of external forces which affect motion, and of these gravity is among the most important, because it acts continuously vertically downward on all bodies and parts of bodies. It pulls airborne objects back to earth, and it must be resisted by postural muscles if an erect stance is to be maintained. Friction and air resistance are also very significant external forces, because they oppose movement of objects. Other external forces include the buoyant force of water, a push by another person, and the upward impetus provided by a trampoline bed or a diving board.

FORCE COMPONENTS

It has already been pointed out that force is a vector quantity having direction as well as magnitude. Sir Isaac Newton discovered that forces act according to the rules of geometry, and much of what was presented in Chapter 2 will now be of use in treatment of forces. When a force is applied in any direction other than vertically or horizontally, it can, through a process called *resolution,* be depicted graphically with vector arrows representing the vertical and horizontal components of the applied force. Resolution was given only brief mention in Chapter 2, where the emphasis was on composition of vectors.

In Figure 3–3, a body is being acted upon by an unbalanced force, designated F_θ, being applied at some angle identified by the Greek letter θ (theta). The angle indicates that the object is being lifted by the force, and the vertical component of the applied force is noted as F_y, since it lies along the y axis of the coordinate system. Similarly, the horizontal component of the force is designated F_x, because it lies along the x axis and is responsible for any horizontal displacement that occurs as a result of this force. The F_x and F_y vectors will always be drawn at right angles to one another, and F_θ will always serve as the hypotenuse of the right triangle.

In putting the shot, punting a football, hitting a lob shot, throwing a

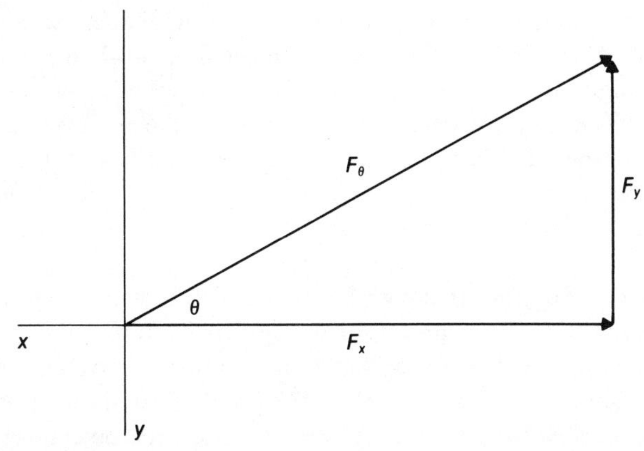

Figure 3-3 Resolution of an applied force F_θ into its F_x and F_y components.

ball, or shooting an arrow, some elevation of the object is needed to offset the ever-present downward pull of gravity. In the shot put, the hand must apply a force F_θ to the shot at some optimal angle θ to the horizontal to provide an F_x horizontal component, as well as an F_y vertical component that will lift the shot sufficiently at release to allow it the necessary time in the air for the horizontal velocity to carry it the desired distance (Figure 3-4). A more detailed discussion of projectiles will be given in the next chapter.

Figure 3-4 Resolution of applied force F_θ during a put at some angle θ.

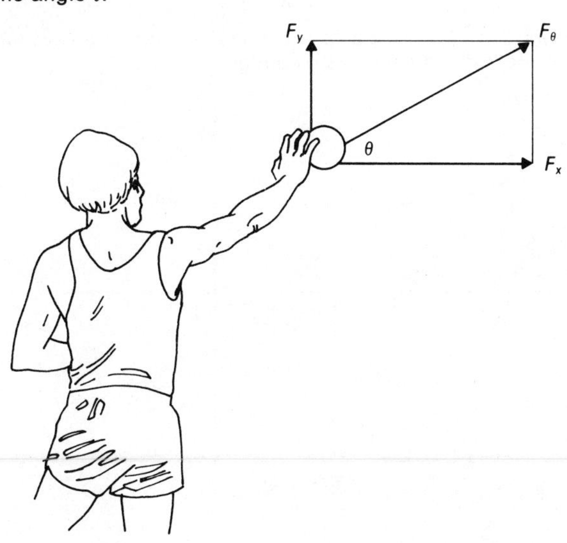

Occasionally two or more forces act simultaneously on a single body to cause it to move. This can be the case internally, as when two muscles act at different angles but on the same segment, or externally, as when two football players block an opponent from different sides at the same instant. The composition of these forces to find a resultant magnitude and direction can be treated in the same manner as the displacement problems presented in Chapter 2.

Example 3–1. Suppose that two forces are acting at once on an object as shown in Figure 3–5. Unbalanced force F_1 exerts 75 lb in a positive direction on a line coinciding with the abscissa. F_2 is an unbalanced force of 50 lb and forms an angle of 30° relative to F_1 at the point of common contact. These component forces will produce a resultant effect on the object, which will be moved at some angle relative to the action lines of the two forces. What is the resultant force and at what angle can movement be presumed to occur?

Method 1. A force vector should always be represented as a *pull* on the point where the force is applied, and the vectors in Figure 3–6 are drawn accordingly. It should be noted that Figures 3–5 and 3–6 are in fact identical. The vectors *AD* and *AB* are drawn carefully to some scale, and a parallelogram is constructed by drawing two more vectors opposite and parallel to *AD* and *AB*. A diagonal line *AC* becomes the resultant effective force acting on the object being moved. The magnitude of this resultant can be measured directly from the graph paper, and the angle relative to either component vector can then be read with a protractor.

Figure 3–5 Placement of two force vectors on a coordinate with the barbs meeting at the point of impact.

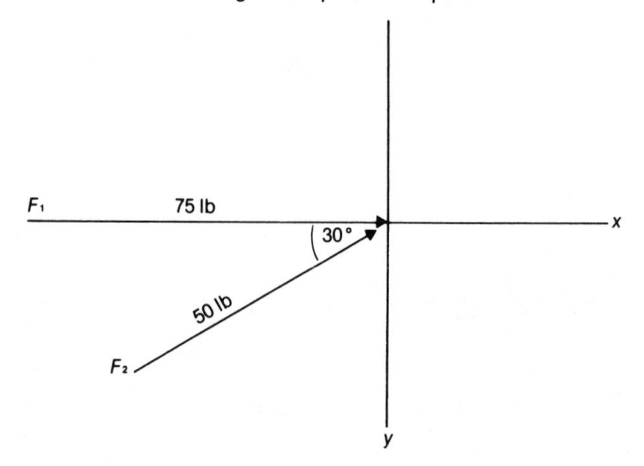

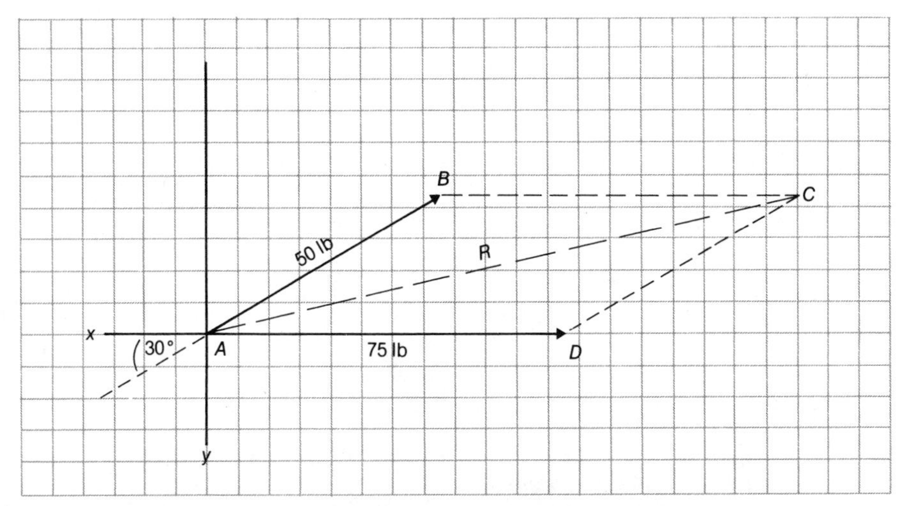

Figure 3-6 The parallelogram method for determining the net effect of two component forces acting on one point.

Method 2. If we wish to set up the problem in the same way we did in Figure 3-6 but without taking the care to draw precise vectors, a right triangle AEB can be created (see Figure 3-7) by simply extending BC over to the y axis, to which it is conveniently perpendicular. Simple trigonometry can now be used to solve for AC in the triangle AEC.

Angle EAB is 60° (that is, 90° − 30°), and the hypotenuse AB is 50 lb. Side EB opposite the known angle is found as follows:

$$\sin 60^\circ = \frac{EB}{50 \text{ lb}}$$

$$.866 = \frac{EB}{50 \text{ lb}}$$

$$EB = 50 \text{ lb} \times .866 = 43.3 \text{ lb}$$

The adjacent side EA is determined as follows:

$$\cos 60^\circ = \frac{EA}{50 \text{ lb}}$$

$$.5 = \frac{EA}{50 \text{ lb}}$$

$$EA = 50 \text{ lb} \times .5 = 25 \text{ lb}$$

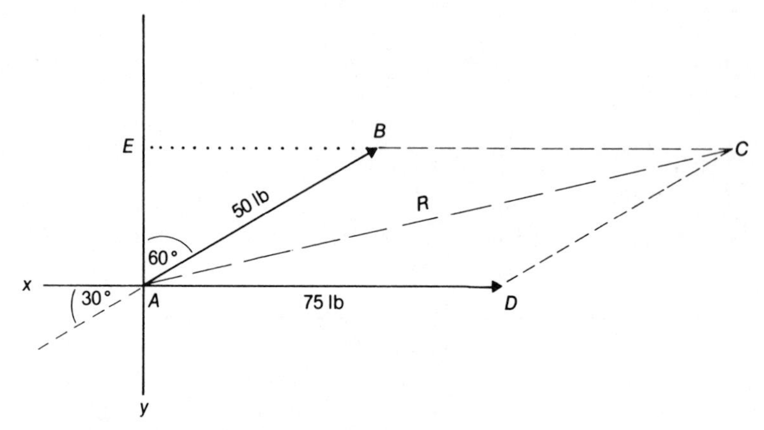

Figure 3-7 Creation of right triangle AEB from a parallelogram laid out on coordinates.

Now, in the larger right triangle *AEC,* side *EA* is 25 lb and *EC* is the sum of *EB* and *BC,* 118.3 lb. Hypotenuse *AC,* the resultant, can be found with the Pythagorean theorem:

$$AC^2 = (25 \text{ lb})^2 + (118.3 \text{ lb})^2$$

$$AC = \sqrt{625 + 13995} \text{ lb} = \sqrt{14620} \text{ lb} = 120.9 \text{ lb}$$

To obtain the angle between resultant *AC* and the 75-lb vector *AD,* any of the trigonometric ratios may be used, as for example:

$$\cos EAC = \frac{\text{adjacent side}}{\text{hypotenuse}}$$

$$= \frac{EA}{AC} = \frac{25 \text{ lb}}{120.9 \text{ lb}}$$

$$= .2068$$

which is approximately the cosine of 78°. Its complement is 12°, the angle between *AC* and *AD.*

Method 3. If Eq. (2-3), the cosine law, is used, there is no need to draw a parallelogram, but it may be helpful to the student to use a schematic arrangement of vectors laid head to tail, as are *AD* and *DC* in Figure 3-7.

$$R^2 = a^2 + b^2 - 2ab\cos\theta$$

$$\begin{aligned} \text{where } a &= \text{the 50 lb force} \\ b &= \text{the 75 lb force} \\ \theta &= \text{angle included between } a \text{ and } b \end{aligned}$$

The angle is an obtuse angle in this problem, and so its cosine will be the negative of the cosine of its supplement.

$$\begin{aligned} R^2 &= (50\text{ lb})^2 + (75\text{ lb})^2 - (2)(50\text{ lb})(75\text{ lb})(-\cos 30^\circ) \\ &= [2500 + 5625 - 7500(-.866)]\text{ lb}^2 \\ &= [8125 - (-6495)]\text{ lb}^2 \\ &= [8125 + 6495]\text{ lb}^2 \\ R &= \sqrt{14620}\text{ lb} = 120.9\text{ lb} \end{aligned}$$

Finally, to calculate the angle between the resultant and the 75-lb vector b, use Eq. (2-6), the sine law:

$$\begin{aligned} \frac{50\text{ lb}}{\sin\theta} &= \frac{120.9\text{ lb}}{\sin 30^\circ} \\ 120.9\text{ lb}\sin\theta &= (50\text{ lb})(.5) \\ \sin\theta &= .2068 \end{aligned}$$

which is approximately the sine of a 12° angle.

MUSCLE FORCES

The actual strength of an isolated muscle in the living human cannot be measured. For most purposes, a resultant torque can be estimated for a group of muscles acting in concert. Similarly, the measurements of attachment angles of muscle tendon on bone cannot be accurately determined. Furthermore, the function of every muscle in every action cannot be always correctly judged, because the muscle's role may vary throughout the performance of a skill. It may act at any given moment as an agonist, an antagonist, a neutralizer, or a stabilizer. Early anatomists pulled on tendons of dissected cadavers to determine the actions of various muscles. Today, the electromyograph is widely used for this purpose, but there are some limitations because the wire attachments are restricting and because only surface muscles are accessible.

A grip dynamometer measures the combined force exertions of the finger and wrist flexor muscles, and other, similar devices are available to test numerous muscle groups. Force can also be estimated by the acceleration given to any known mass by the efforts of a specified muscle group.

While actual forces for a single muscle, as we have just said, cannot be measured with any precision, there is some value to be derived from working with hypothetical forces to show the relationship of component forces. Figure 3-8 is a free body diagram of an arm being flexed at the elbow by one elbow flexor muscle. The angle of concern here is not that at the joint but rather the one formed between the muscle tendon and the bone to which it attaches.

Assume that the muscle is concentrically contracting with a force of 100 lb and that at this point of flexion the angle of attachment of the muscle tendon to the bone being moved is 50°. With these data, the horizontal and vertical force components can be computed. In this case, F_x is called the *compression* (or *stabilizing*) component of the force, because it represents the amount of muscular force being used to bring the head of the moving bone into contact with the end of the more proximal bone, thereby helping to stabilize the joint. The F_y component is called the *rotatory* component, since it represents the amount of force available to rotate the forearm around the elbow joint. Since Figure 3-8 depicts a right triangle when the vectors are laid head to tail, the components are obtained with trigonometry as follows:

Figure 3-8 A free body diagram showing the component forces when a known muscle force is pulling at a given angle at a particular instant of elbow flexion.

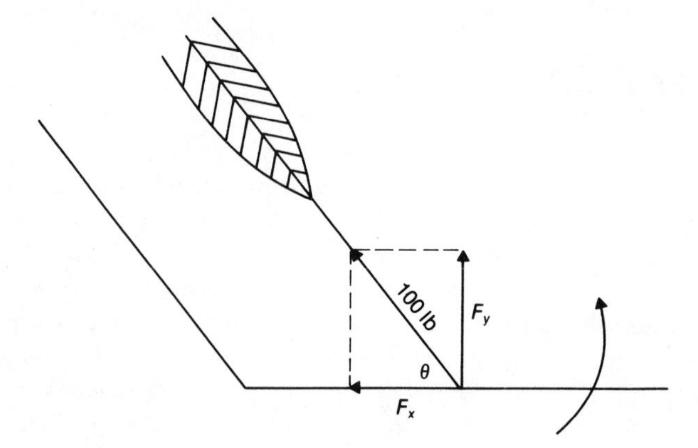

$$\sin 50^\circ = \frac{\text{opposite side}}{\text{hypotenuse}}$$

$$.766 = \frac{F_y}{100 \text{ lb}}$$

$$F_y = 76.6 \text{ lb} \qquad \text{rotational force at } \textit{this} \text{ angle}$$

$$\cos 50^\circ = \frac{\text{adjacent side}}{\text{hypotenuse}}$$

$$.6428 = \frac{F_x}{100 \text{ lb}}$$

$$F_x = 64.28 \text{ lb} \qquad \text{compressive force at this angle}$$

At an attachment angle of 45°, sine and cosine are identical, and so the two component forces are the same. When the angle is greater than 45°, the rotatory component is the larger. When the angle is less than 45°, it is the compression component that is the larger.

This entire discussion is rather academic and serves only to point out the role played by attachment angles throughout the range of motion of any movable segment. Even if the angles could be measured, neither the coach nor the athlete could do anything to change internal muscle attachment angles. If an individual's single muscle force could be determined with any accuracy, it would be effective only for a particular angle and even only for a particular speed, because the faster the contraction, the less the force that is available. All in all, it is a complicated matter.

PRESSURE

Till now, force has been considered from the standpoint of its potential to cause motion. Another aspect of concern in many sports is the reduction of injury resulting from blows or falls, and the area over which an impact force is distributed has a direct bearing upon the resulting pain or injury. This distribution of force per unit of area is called *pressure*.

$$\text{Pressure} = \frac{\text{total force}}{\text{area of application}}$$

The units are in pounds per square inch (lb/in^2) or newtons per square meter (N/m^2). The force is that component acting perpendicularly to the surface.

One of the prime principles related to the minimizing of injuries in falling is that the force of impact should be received over as large an area of body surface as is possible under the circumstances. For example, athletes can expect more harm if they land on an elbow than if they land on thigh and trunk. Learning how to fall properly should be part of every athlete's training in those sports where a runner may be tripped or knocked down or where a fall is the final part of the performance, as in pole vaulting. The old vaudeville slapstick comedian was often a master of the pratfall, and Hollywood stunt men would have short careers indeed if they could not fall or tumble without injury. Students of various martial arts and of mime spend considerable time in learning falling techniques.

Properly designed sports equipment can appreciably reduce pressure in contact sports. Football shoulder pads distribute the forces of blocking and tackling over a broad area. This is also the purpose of helmets in hockey, lacrosse, baseball, boxing, and cycling. A large boxing glove allows the force of a punch to be spread over a broader facial surface than would a smaller-sized glove (or bare knuckles) and so is desirable for beginners.

Where infliction of pain or injury *is sought,* blows should be given with the smallest possible impact area. A slap to the face will certainly sting, but a similar blow made with the edge of the open hand, karate style, will concentrate the force and can cause severe injury. For the same reason, riot police are taught to jab with the end of a billy club rather than to swing it. An ordinary magazine would not normally be considered useful as a weapon of defense, but when it is tightly rolled up, a jab made with the end can inflict injury to an attacker.

An ice skater's blade exerts great pressure on the ice because the skater's full weight is transmitted to the ice via a very small blade-to-ice contact area. This pressure is enough to actually melt the ice so that the blade glides over a film of water.

One must take care to use only well-designed and well-fitting protective equipment, since it is possible that instead of distributing force from a blow, poor equipment might redirect the force to an unprotected part of the body. It is also possible that such equipment as hockey face masks might not really distribute much of the impact force but instead only reduce peripheral vision and make the player vulnerable to other injuries.

A notable exception to the principle that a small impact area during a fall will cause injury is in the sport of diving, where a small, clean, nearly splash-free entry into the water is very desirable both to avoid a stinging pain and to achieve a high score from the judges. Where the entry has to be shallow, as in a lifesaver's jump, a broad body area of contact accompanied by a well-timed downward thrust of arms and legs will allow a swimmer to jump into the water without submerging the head. From low jumping heights, this type of entry causes only minor discomfort in achieving the desired purpose.

FRICTION

Everyone has some idea of the meaning of *friction,* but many consider it to be something undesirable. It is often associated with heat buildup, as would happen in unlubricated bearings. The friction of rubbing two sticks together can be used to start campfires, and we warm our hands in the winter by rubbing them together. Reduction of friction is important in sports such as skiing, swimming, and roller-skating, where a person has to glide over a surface or move through a fluid. Within the body there are numerous bursae located strategically to reduce friction where tendons might rub over other tissue. Synovial fluid is secreted to lubricate joints, and the ends of the articulating bones are smooth and covered with slippery hyaline cartilage.

Friction may be defined as a force that resists or opposes the movement of one body over another. It acts parallel to the two surfaces and in a direction opposite to that of the motion. Where either or both of the two surfaces are oval, the frictional force acts tangentially at the point of contact.

When you think about forces, you probably picture something actively pushing or pulling something else and somehow do not envision friction as such a force. This may be because friction is passive in the sense that it is always a reaction to some other force. When no motion is being attempted, friction does not exist. A book lying on a table experiences no frictional force as long as it is motionless.

Runners depend upon friction to prevent slipping at the start and during every propulsive step, and without it they would not get the necessary ground reaction to move forward. A quarterback requires friction between football and fingers during a pass attempt. A man making a sharp cut in basketball must have adequate friction between his shoes and the floor. He will slip and fall if the force he exerts in any direction with his feet is greater than the frictional force in reaction.

In general, the magnitude of friction depends upon:

1. The types of materials of which the contacting bodies are made. One would expect a rubber-soled shoe to have better traction on a wooden floor than would a leather-soled shoe.
2. The relative roughness or smoothness of the contacting surfaces and the presence of moisture, lubricants, or dust. Cleats are used on turf to help the player get a better grip on the surface, particularly during stops, starts, and turns. Unfortunately, if the foot is too firmly fixed at the moment of a tackle or a block, serious injury to the knee may result.
3. The normal (perpendicular) force pressing the surfaces together.

On a cold day, lightly rubbing the palms together will warm the hands, but pressing harder will increase the heat produced by the friction.

4. The state of motion or rest between the surfaces. This will be explained in the following discussion.

There are four categories of friction—static, kinetic, rolling, and fluid. While there are common elements, each will be discussed separately.

Static Friction

Static friction, sometimes called *starting friction,* is that which is encountered when one attempts to put some stationary object into motion over a supporting or an adjacent surface. The initial effort that is exerted to push a heavy filing cabinet across an office floor is opposed by static friction until the pushing force is great enough that the limit of this friction is reached and the cabinet begins to move. The amount of force required will depend upon the weight of the cabinet, the smoothness of its bottom, the kind of floor upon which it rests, and the direction and point of application of the force. A sharp change of direction will not make an athlete slip as long as the limit of starting friction has not been exceeded between the athlete's shoes and the floor.

The coefficient of static friction between two contacting surfaces is simply the ratio of (1) the force required to overcome friction to (2) the normal force, usually the weight, pressing the surfaces together when the surfaces are at rest relative to one another. Thus,

$$\text{Coefficient of static friction} = \frac{\text{moving force}}{\text{pressing force}}$$

The smaller the value of the coefficient, the less the friction between the surfaces. Symbolically this coefficient is usually expressed as

$$\mu_s = \frac{F_s}{N} \tag{3-1}$$

where F_s = force needed to begin motion

N = force acting perpendicularly to press the surfaces together

μ_s (the Greek letter mu) = symbol for the coefficient of friction

Another approach to the measurement of the coefficient is only occasionally feasible. If the supporting surface can be inclined while supporting an object, the point at which the inclination is just enough to cause the object to begin sliding is where the coefficient of friction is equal to the tangent of the angle formed by the incline with the horizontal. Bunn refers to this as the *limiting angle.*[2]

$$\mu = \tan\theta \qquad (3\text{–}2)$$

Kinetic Friction

Kinetic friction, sometimes called *sliding friction,* is the ratio of the force required to overcome friction between two surfaces to the normal force pressing the surfaces together when one surface is sliding over the other at a constant speed. Most of us know from experience that it is somewhat easier to keep an object moving once we get it into motion, and so kinetic friction is always somewhat less than static friction. The ratio equation is the same as Eq. (3–1) but with a change in subscript symbols and with the understanding that motion already exists:

$$\mu_k = \frac{F_k}{N} \qquad (3\text{–}3)$$

where F_k = force needed to keep one surface moving over another at a constant speed

N = the normal pressing force

μ_k = coefficient of kinetic friction

Examples of kinetic friction in sports are fairly limited; coming to mind are sledding, skiing, and ice skating, which involve a sliding over snow or ice.

To summarize the coefficients of static and kinetic friction, static friction resists the beginning of motion and is of greater concern to coaches and athletes than is kinetic friction, which is the resistance to maintaining motion. Notice that there is no mention of area of surface contact in either Eq. (3–1) or Eq. (3–3), because the coefficients are independent of such area. It is important to remember that any coefficient is valid only for the two particular materials and surfaces used in the ratio. For example, adding moisture will change the coefficient. The caliper brakes of a bicycle work less well on a rainy day, and the cyclist must allow many times the normal distance to stop.

[2] Bunn, *Scientific Principles,* p. 66.

If the coefficient of friction is known for two materials, the force required to just begin motion can be determined by rearranging Eq. (3–1) so that $F_s = \mu_s N$. It is obvious here that the force is proportional to the normal pressing force.

Before friction can exist to oppose motion, there must first be a lateral force acting or a force component which is directed parallel to the surface. A downward pressing force, as when a person stands in place, does not elicit a frictional reaction, because the force is entirely perpendicular to the surface, with no component in any possible direction of motion. Even when a lateral component is present, there will be no motion until the limiting friction value is reached. Until then, the frictional force always equals the applied force. Tricker and Tricker state that friction is "a self-adjusting force which just balances any force that may be applied and so prevents sliding taking place."[3]

The practicing coach, teacher, or athlete learns empirically all that is really necessary about friction as it relates to his or her sports specialty. If several players have fallen during practice, the basketball coach will quickly suspect either that the floor is slippery or dirty or that it is time to buy new shoes for the players. The short-term solution, the immediate remedy, is to have the floor mopped or to advise the players not to cut quite so sharply. Where the problem does not have such a ready answer, consultation with gym floor specialists might be necessary. But a mathematical determination of friction coefficients is important only to manufacturers of sports equipment or to researchers working on the reduction or improvement of friction characteristics of sports equipment and playing surfaces.

Rolling Friction

Rolling friction is encountered by a ball moving over the ground and is usually the result of one or both surfaces deforming during contact. A hard, smooth ball on a hard, smooth surface will roll quite easily. Indoors, rolling friction is present in such sports as bowling, billiards, and roller skating. Conditions outdoors are much more variable in that a ball's behavior is partly influenced by the height of the grass, the hardness of the ground, the smoothness of the surface, and the presence of moisture. Golf comes to mind as one sport where these factors must be accounted for throughout a game. Other examples include baseball, lacrosse, soccer, field hockey, lawn bowling, and croquet. Balls from each of these sports, when rolled on identical surfaces, will of course have different friction coefficients because of their individual differences in surface texture, hardness, weight, and diameter.

[3] R. A. R. Tricker and B. J. K. Tricker, *The Science of Movement* (New York: American Elsevier Publishing Company, Inc., 1967), p. 21.

Fluid Friction

Fluid friction is a difficult topic to present in a meaningful manner to athletic practitioners, because we are dealing with the obvious, more simple aspects of friction—an object's surface smoothness, its shape, and its velocity. Any study at the next level would require a knowledge of aerodynamic or hydrodynamic engineering. Fluid friction results from the movement of an object through either air or water. It does not matter actually whether the body is moving through the fluid or the body is still and the fluid is flowing past it, as it might in a wind tunnel test. The fluid is disturbed by the object, which exerts a force on it and pushes it aside or ahead. Thus work is being done and the motion of the body and of the fluid is being changed in some manner. When a large truck passes a car on the highway, the air being pushed aside is often able to affect the car and require a steering correction by its driver.

The degree of disturbance of the fluid depends upon, among other things, the shape of the object and its velocity. Streamlining an object can reduce drag by allowing a smooth fluid flow over the object from front to rear with a minimum of energy-consuming turbulence formed in the object's wake.

Air resistance is a form of fluid friction that affects athletes in some sports more than in others. In sports that feature rapid movement over some distance, air resistance must be taken into account. Hence, those in track, skiing, speed skating, and cycling are greatly affected by air resistance, whereas weight lifters, bowlers, squash players, and fencers do not consider it at all. The air not only acts on the athlete but may also act upon anything projected, such as a tennis ball, football, or javelin.

There are rather few ways in which an athlete can reduce air resistance or its effects. Form-fitting, smoothly textured clothing might work, or as in cycling or downhill skiing, streamlining the body position may be helpful. But there are limits to the configurations that can be assumed by the body. A man running a 200-meter dash simply must sprint as fast as he can even though air resistance increases as the square of his velocity. Cycle racers know that the lead cyclist assumes the greatest resistance and that tailgating him or her places the trailing person in a pocket of reduced resistance, and so it is not always an advantage to be the leader in the early stages of a race.

One area of study related to air resistance that has received a good deal of attention over the years is the flight of balls, both spinning and nonspinning. Baseball pitchers long ago learned that spitballs or balls with scuffed surfaces, both illegal, behave differently from normal baseballs. Manufacturers of all types of balls have had a natural concern for producing balls that would fly farther and with more consistency than those of competitors. Thus trial-and-error studies have been mixed with scientific ones.

Wind tunnel tests provide the best means to study the effects of air on a ball. Such tests have shown that air resistance is related to a ball's linear velocity, shape, and surface characteristics. Of greater interest has been the behavior of spinning balls, which requires consideration of the rate of spin as well as of the linear velocity and the condition of the ball surface.

As a ball spins while moving in a linear path, it pulls around with it a thin layer of air, a boundary layer, which interacts with the layers of adjacent air. In the case of topspin, there is a lower net velocity at the top, where the boundary layer is moving in a direction opposite that of the air it meets (Figure 3–9). At the bottom of the ball, the boundary layer is moving in the same direction as the air it encounters, and this results in a higher net velocity. As the eighteenth-century Swiss mathematician Daniel Bernoulli observed, pressure in a fluid decreases as velocity increases, and so the air pressure difference between the top and the bottom causes the ball to curve downward toward the lower pressure. For the same reason, balls with back spin will tend to rise and balls spinning around a vertical axis will curve in the direction of the spin. This curved flight resulting from the combined effects of spin and air resistance is sometimes called the Magnus effect.

Engineers working for golf ball manufacturers have experimented with the size, shape, and depth of and the number of dimples on a golf ball in order to improve flight characteristics in terms of both distance and stability. It was noticed many years ago that the original smooth golf balls did not travel as far when new as they did after they became scuffed. The apparent reason for this is that the boundary layer on a smooth ball separates early and this results in vortexes and a lowered pressure behind the ball. Rough-surfaced balls delay the separation of the boundary layer and experience less drag. According to Walker, dimpled golf balls travel up to four times farther than smooth balls.[4]

Two terms that are commonly used in discussions of air resistance are the opposing forces of *lift* and *drag*. For sports objects, such as discuses and javelins, which have airfoil characteristics, there is a lift force perpendicular to the direction of the object's motion and opposed by the weight of the object. The airfoil shape and the angle at which the body meets the air (the angle of attack) are two determinants of the lift magnitude. A drag force is one which opposes propulsive efforts to move through a fluid, and its magnitude depends upon the shape and the velocity of the object. The shape determines the amount of turbulence in the form of eddies and whirls produced behind the moving body. A streamlined object has a shape that reduces eddies because of the smooth laminar flow of fluid around it, and it loses less kinetic energy to turbulence.

The principles relating to air resistance also apply to water resistance. Over many centuries, sailors and shipbuilders have sought ways to in-

[4] Jearl Walker, "The Amateur Scientist," *Scientific American,* April 1979, p. 180.

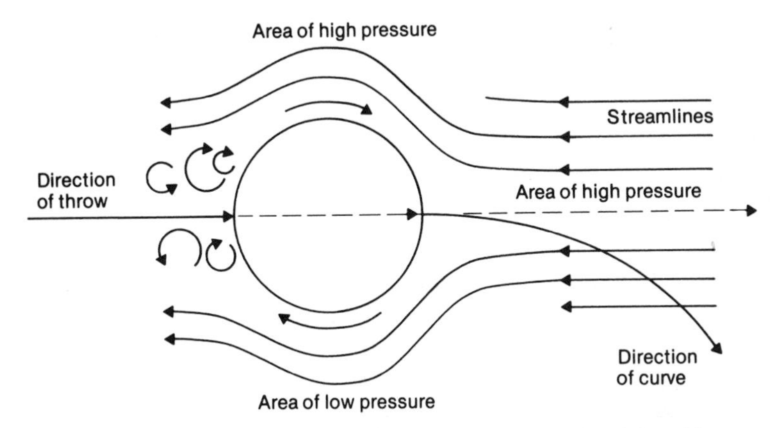

Figure 3-9 A ball with topspin moving toward the right will curve downward because of the greater pressure on the top of the ball.

crease the speed of vessels. Enormous sums have been spent in efforts to improve the hull designs of world-class racing yachts and of naval vessels. However, marine architects and wealthy yacht racers are not the only ones who are concerned with efficient movement through the water. Swimmers for years have tried to improve their competitive times through such varied approaches as head and body shaving, using skintight racing suits, and employing different starting and stroking techniques. These efforts are typically directed toward either reducing the various resistances affecting swimmers or improving the propulsive force-producing capabilities of the arms and legs.

Resistance encountered by a body moving through water is a function of (1) the cross-sectional area that the body presents at right angles to the flow, (2) the friction of the skin or clothing, (3) the turbulence in the rear consisting of energy-consuming eddies, and (4) the velocity of the body. Unlike a human swimmer, a fish presents a streamlined shape to the water it meets. The more smoothly the water can flow around the body, the less drag there is to retard movement. But as is true in land locomotion, resistance is needed for effective propulsion through water.

The human body propelled through the water by any of the four competitive strokes presents a complex problem in hydrodynamics. Probably more is known about the swimming mechanics of fish than about humans, but in recent years there has been intensive study of swimming by biomechanicians worldwide, and a number of international seminars on swimming mechanics have already been held. The dramatic improvement of swimming times over the past twenty years is directly due to the increased participation in age-group swimming and more intensive training. Secondarily, some rules changes regarding starts and turns have been a factor in lower times, as has the engineering of faster pools and lane lines with improved wave-damping capabilities.

BUOYANCY

A force related to aquatic activities is that of *buoyancy,* which may be defined as the upward force of a liquid on a wholly or partly immersed body. The principles of buoyancy are of some limited value to teachers of beginning swimming, who might normally introduce floating skills early in the program.

Archimedes (287–212 B.C.) stated that a body immersed in a fluid is buoyed up by a force equal to the weight of the fluid that is displaced by the body. In other words, an immersed body loses weight equal to the weight of the displaced water.

Some lean, muscular people, usually men, are *sinkers* and cannot be taught to float because of the ratio of body density to water density. This ratio is the *specific gravity* of a body and is found by dividing body weight by displaced water weight. A ratio above 1.0 indicates that the displaced fluid weighs less than the swimmer and so the swimmer will sink. A ratio less than 1.0 means that floating will occur. Women typically are more buoyant than men because they tend to have more fat, which is not as dense as muscle or compact bone.

If a particular man has a specific gravity that will allow him to float, he may still encounter difficulties in achieving a desirable position. A prone float has limited practical value aside from being easy to learn and being a necessary first step in learning the crawl stroke. Most swimmers at some time are introduced to the supine float with the face out of the water on the premise that this allows the swimmer to rest while breathing continuously. A perfectly horizontal supine float is not usually possible, but the majority of beginners can be taught to float comfortably with the body at some angle between the horizontal and the vertical. Provided that the swimmer does not assist the floating effort with hand or leg movements, the floating position that a particular individual can assume is determined by the relative locations of the body's centers of gravity and buoyancy.

The *center of buoyancy* of an immersed body is that point through which the upward force of buoyancy acts. Its location is somewhat affected by the volume of air taken in and held by the lungs at any given moment. Figure 3–10 illustrates that the center of buoyancy is generally more cephalad, that is, closer to the head, than is the mass center. The closer together these two points are, the more nearly horizontal the floating position can be. The farther apart the two centers are, the more rapidly the legs will submerge because of the turning moment which acts on the legs.

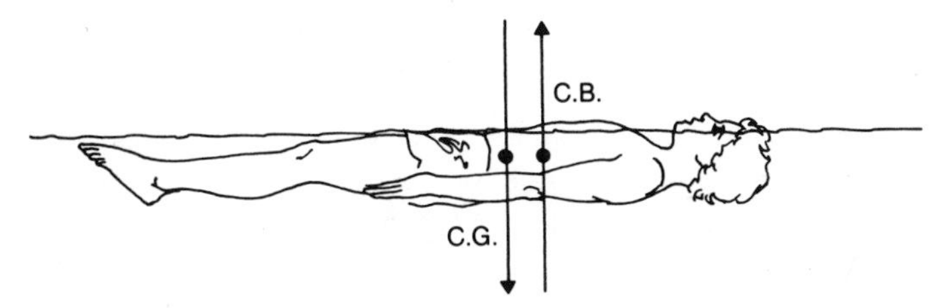

Figure 3-10 A horizontal supine float is possible if the centers of gravity and buoyancy are close together.

By manipulation of the limbs, the distance between the centers of gravity and buoyancy can be altered. The center of gravity can be raised by extending the arms overhead and also by flexing the legs at the hips and knees. This should enhance the supine horizontal float capability.

When the final floating angle has been achieved, the legs will have stopped rotating downward (in some cases upward) and the center of buoyancy will be in a vertical line with and usually above the center of gravity (Figure 3-11). The teacher must understand how difficult it is for a beginner to be totally relaxed long enough to achieve the desired arched-back, hyperextended-neck position in a wavy pool. A gentle flutter kick along with a sculling action of the hands requires little energy or skill but makes a horizontal float possible and more practical than a completely motionless float.

Figure 3-11 When a stationary floating position is achieved, the center of buoyancy will be in a vertical line with the center of gravity.

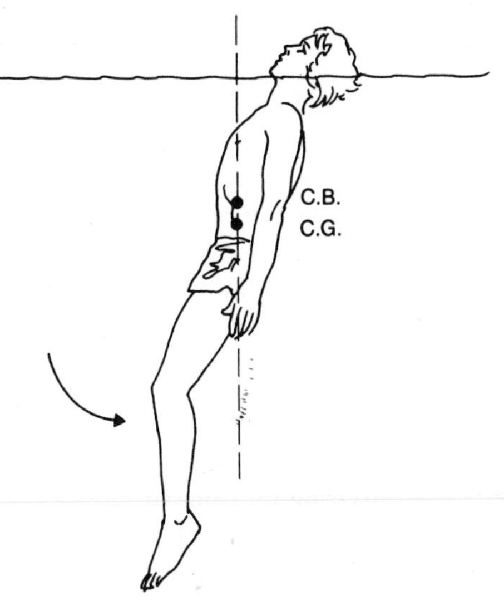

SUMMARY AND DISCUSSION

Forces are acting upon us continually, and we exert forces constantly. Every time you turn a page of this book you apply an external force and your muscles provide an internal force in your arm, wrist, and fingers. Nothing moves or stops moving unless an unbalanced force acts upon it. *How* the object that is acted upon moves depends on how much force is applied, the point of application, and the direction of application.

Isotonic contractions (which cause a change in muscle length) are termed concentric if the muscle shortens, while so-called lengthening contractions are called eccentric. Students are frequently confused about the role played by a muscle during particular movements, and this is largely because they assume that if a limb is being extended, then the extensor muscles must be involved. This is true only if the movement is not being done against the force of gravity or some similar external resistance or if the movement requires the limb to be moved at an acceleration higher than 32 ft/s^2. Otherwise, the controlled lowering of a weight at less than gravitational acceleration requires the use of those muscles which originally lifted the weight, be they extensors or flexors, as prime movers.

Isometric contractions involve a maximal force exertion without any resulting movement. Often this entails the deliberate pitting of one set of muscles against another, as when the palms are placed against each other and an all-out effort is made for several seconds followed by a brief rest and repetition of the exercise. This example would tend to develop the muscles that horizontally flex the humerus, notably the pectoralis major and the anterior deltoid, and is therefore an exercise to develop the anterior chest. Several commercial exercise programs are built around the use of isometric exercises, and while these have merit, they do not exercise the muscles through a full range of motion. Their major advantages are in not requiring the use of much equipment and in their suitability to the bedridden.

Forces directed at any angle between the horizontal and the vertical have two components to consider. A horizontally applied force has no vertical component, and a vertically applied force has no horizontal component. But when you throw a ball in the usual way, the point of aim is high enough to give the ball time to move the desired distance. This elevation requires the applied force to have a vertical component, and of course a horizontal component is needed to achieve a distance goal. When these components are related to internal forces of muscles, the horizontal vector is called the *compression* component and the vertical vector is termed the *rotatory* component. These determine how much of the actual muscle force is used to move the limb and how much is "wasted" in pressing the joint surfaces together. When the tendon attachment angle exceeds 90°, the compressing tendency changes to a dislocating tendency, although actual dislocation does not ever result.

Pressure is the force per unit of area of one body on another. Low pressure is needed to avoid pain or injury in falls or from blows. The key is to spread the impact force over as large an area as possible. Athletes in gym shoes do no damage to a gym floor, but consider the force per unit of area when track spikes or metal cleats are worn! The floor surface could be punctured by the transmission of an athlete's body weight to the floor via a small number of contact points in the form of cleats or spikes.

It should be clear by now that friction is an absolute necessity in every sport. Even the skier who coats his or her skis to reduce friction finds that friction is needed between the edges and the snow to make turns, and between the poles and the snow for control. Those who have used towrope lifts know that the rope must be allowed to slip during the initial grip to prevent the shoulder wrench that comes with an immediate tight grip. During such slipping, the hand will burn from the rope friction unless a glove is worn.

A roofer can walk up and down incredibly steep slopes in applying shingles or making repairs, but even such an experienced person cannot walk when the roof pitch exceeds a given angle determined by the coefficient of friction between shoe and the shingle. This coefficient depends on the nature of the contacting materials—that is, on what the materials are and whether their surfaces are rough or smooth—and on the normal force that presses the surfaces together. Note that the roofer's body weight is not the normal force, because only that component of the weight acting perpendicularly to the roof surface is considered.

The static friction between two objects is always somewhat greater than the kinetic friction, and as one might expect, rolling friction is considerably less than either static or kinetic friction.

Resistance of a body moving through a fluid (air or water) is related to the body's shape, surface texture, and velocity, all of which contribute to the degree of disturbance of the laminar flow of the fluid around the body. The faster, the more blunt, the broader, and the more roughly surfaced the object is, the greater will be the drag. In sports, the objects that offer the most resistance to movement through the air are parachutes and badminton shuttlecocks.

The behavior of objects moving through the air as projectiles, both spinning and nonspinning, has held the interest of physicists and coaches for a long time. For years there were scientists who insisted that a pitched baseball did not really curve, and that any apparent curves were just illusions. That spinning balls do curve has now been well established, but research continues into the conditions associated with such curving.

Pressure differences result as a spinning ball meets oncoming air. The lower pressure on the side of the ball that moves *with* the air causes the ball to deviate in that direction. Thus, backspinning balls tend to rise out of the normal parabolic path, and topspinning balls tend to drop faster than they would with no spin.

Spin also tends to stabilize an object in flight. Footballs, discuses, and Frisbees are typically released with spin for that reason. Releasing the object with the "nose up," that is, with an angle of attack, provides a lift component, which, together with an airfoil shape, prolongs the object's flight by pushing downward on the air, which gives an opposing lift reaction perpendicular to the direction of motion.

Forces, then, both resistive and propulsive, are among the major concerns of biomechanicians, athletes, coaches, and teachers. Success is often closely linked to the discovery of the right combination and balance of forces.

Problems for the Student

1. A shot-putter applies 80 lb of unbalanced force to a shot at an angle of 41° to the horizontal. Find the two component forces.
2. Two forces are acting on a body, one 30 lb and the other 25 lb. If the angle between these forces where they meet on the body is 27°, what is the resultant force and at what angle is the resultant relative to the 30-lb vector? Solve by two different methods.
3. If a brachioradialis muscle is pulling on a radius at an angle of 9° with a force of 80 lb, how much of that applied force is available to rotate the radius around the elbow axis? Assume that the humerus is fully stabilized.
4. Which principle explains why snow shoes keep the wearer from sinking into deep snow?
5. Given a coefficient of friction of 0.6 between two surfaces, what force will be needed to get one surface moving over the other if a normal force of 50 lb is acting to press the surfaces together?

SUGGESTED READINGS

ARIEL, GIDEON, "Computerized Biomechanical Analysis of Human Performance," in J. L. Bleustein, ed., *Mechanics and Sports* (New York: American Society of Mechanical Engineers, 1973), pp. 267–75.

BRIGGS, L. J., "Effect of Spin and Speed on the Lateral Deflection [curve] of a Baseball; and the Magnus Effect for Smooth Spheres," *American Journal of Physics,* vol. 27, no. 8, 1959, p. 589-96.

CAMPNEY, H. K., and R. W. WEHR, "An Interpretation of the Strength Differences Associated with Varying Angles of Pull," *Research Quarterly,* vol. 36, no. 4, December 1965, pp. 403–12.

CLARYS, J. P., and J. JISKOOT, "Total Resistance of Selected Body Positions in the Front Crawl," in *Swimming II,* J. P. Clarys and L. Lewille, eds. (Baltimore: University Park Press, 1975), pp. 110–17.

COCHRAN, A., and J. STOBBS, *The Search for the Perfect Swing* (Philadelphia: J. B. Lippincott Company, 1968).

DYSON, GEOFFREY, *The Mechanics of Athletics* (London: University of London Press Ltd, 1967).

FARIA, I. E., and P. R. CAVANAGH, *Physiology and Biomechanics of Cycling* (New York: John Wiley & Sons, Inc., 1978).

HAY, JAMES, *The Biomechanics of Sports Techniques,* 2nd ed. (Englewood Cliffs, N.J.: Prentice-Hall, Inc., 1978).

HERTEL, HEINRICH, *Structure-Form-Movement* (New York: Reinhold Publishing Corp., 1966).

KARPOVICH, P. V., "Water Resistance in Swimming," *Research Quarterly,* vol. 4, October 1933.

PINE, JEROME, *Contemporary Physics* (New York: McGraw-Hill Book Company, 1972).

ROEBUCK, J. A., JR., K. H. E. KROEMER, and W. G. THOMSON, *Engineering Anthropometry Methods* (New York: John Wiley & Sons, Inc., 1975).

SELIN, C., "An Analysis of the Aerodynamics of Pitched Baseballs," *Research Quarterly,* vol. 30, May 1959.

SHANEBROOK, J. R., and R. D. JASZCZAK, "Aerodynamic Drag Analysis of Runners," *Medicine and Science in Sports,* vol. 8, no. 1, Spring 1976, pp. 43–46.

SHELTON, JAY, "The Physics of Frisbee Flight," in Stancil E. D. Johnson, *Frisbee* (New York: Workman Publishing Co., 1975).

VERWIEBE, F. L., "Does a Ball Curve?" *American Journal of Physics,* vol. 10, 1942, pp. 119–20.

WALKER, JEARL, "The Amateur Scientist," *Scientific American,* April 1979, p. 180.

WATTS, R. C., and E. SAWYER, "Aerodynamics of a Knuckleball," *American Journal of Physics,* vol. 43, no. 11, November 1975, pp. 960–63.

WHITT, F. R., and DAVID G. WILSON, *Bicycling Science: Ergonomics and Mechanics* (Cambridge, Mass.: MIT Press, 1974).

WILLIAMS, J. G. P., and A. C. SCOTT, eds., *Rowing: A Scientific Approach* (Cranbury, N.J.: A. S. Barnes & Company, Inc., 1967).

CHAPTER 4

Linear motion

There are basically two categories of motion—linear and angular. Since angular motion will be discussed in Chapter 5, it will suffice here to define it as motion around an axis. Linear motion, sometimes called *translatory motion,* exists when all parts of a body move simultaneously in the same direction for the same distance, as for example when a boy coasts down a hill on his sled. This definition can be unnecessarily restrictive when it is applied to sports actions, in which the real concern usually is for the displacement of the mass center of, say, a runner who is traveling down the track in a straight line. It serves no useful purpose to suggest that, because the runner's arms and legs are moving in an angular manner, such movements thereby rule out the classification of running as a form of linear motion. Bodies in motion will be treated as though they were rigid and as though each were a particle whose mass is the actual mass of the whole body concentrated at the center of gravity.

Two types of linear motion can be distinguished. *Rectilinear* motion exists when a body moves in an essentially straight line from one point to another. A ball dropped from the hands to the floor undergoes rectilinear motion. If the ball were to be rolled in a straight line along the floor, its motion would be linear while at the same time its rotation would be classified as angular motion.

If the ball were to be thrown some distance, the force of gravity would act to cause the ball to follow a curved path known as a *parabola,* and its motion would be described as *curvilinear,* that is, motion which is diverted from a straight-line path into one that is curved.

The linear motion of a body can be uniform or nonuniform. In other words, a body may move with either a constant or a varying velocity. Motion is considered uniform when equal distances are covered in equal times, and nonuniform when unequal distances are covered in equal times (Figure 4–1).

For the most part, we will be dealing only with problems of uniform motion, or at least, a condition of uniform motion will be assumed. Admittedly, nonuniform motion is far more common in sports, because athletes frequently change speed or direction, but solving problems in such cases can be quite complex. At any rate, the assumption will be made of constancy of athletes' accelerations and velocities, and in some cases an average acceleration or velocity will be used.

NEWTON'S LAWS OF MOTION

In the year 1687, Sir Isaac Newton published his famous treatise *Philosophiae Naturalis Principia Mathematica,* in which he expressed the laws of motion. These laws, along with Newton's discovery that forces act according to the rules of geometry, have formed the basis for modern mechanics and, hence, sports biomechanics. Galileo Galilei (1564–1642) had done much early work on the principles of movement, which set the stage for Newton's contributions

Figure 4–1 In uniform motion, equal distances are covered in equal increments of time, while in nonuniform motion unequal distances are covered in equal time units.

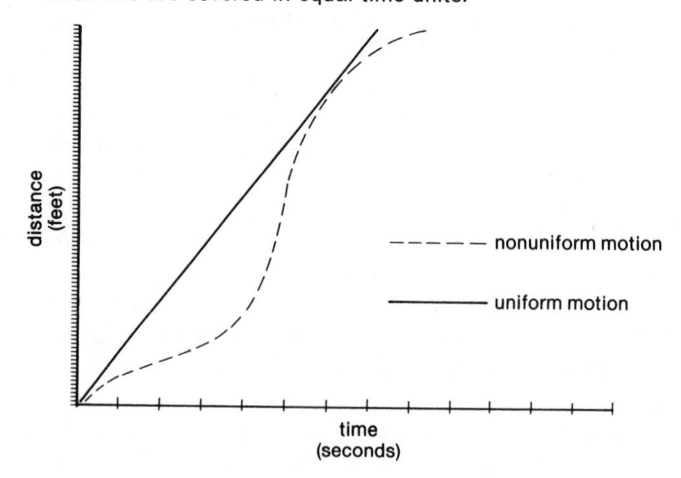

and publication. As paraphrased below, the laws apply to linear motion, but in the next chapter they will be restated in terms of angular motion.

Law 1: The Law of Inertia

A body will remain in its state of rest or of constant linear velocity unless it is acted upon by some external unbalanced force.

In other words, a body will remain in a state of *equilibrium* either in the absence of external forces or when the net sum of all the external forces acting on the body equals zero. A thrown ball will not go on forever in a straight line with uniform velocity, because of gravity and air resistance, two external forces that slow the ball and pull it downward. In the same manner a rolling ball will eventually come to rest because of frictional force, which always opposes the motion of one body over another.

Inertia, a word derived from the Latin for *idleness,* may be defined as the tendency of a body to resist any change in its state of linear motion. Another definition is that inertia is a property of matter that requires some force exertion to change a body's condition of motion. Rest is a special state of motion, and a body will remain at rest forever unless an external unbalanced force acts upon it.

Inertia is said to be proportional to mass, the quantity of matter in a body, and so the heavier the object, the more the force required to set it into motion or to bring it to a stop. It requires greater force to tackle a football fullback than it does to stop a lighter halfback moving at the same speed. Note that a body's inertia is only one factor in the determination of resistance to movement. Frictional forces and stability factors must also be considered.

Law 2: The Law of Acceleration

When an unbalanced force is applied to it, a body experiences an acceleration which is directly proportional to the unbalanced force, is in the same direction as the unbalanced force, and is inversely proportional to the mass of the body.

Acceleration is a vector quantity and is defined as the rate of change of velocity, that is, a change in either speed or direction of movement. It is expressed as some number of feet per second per second (ft/s^2) or of meters per second per second (m/s^2) in a particular direction. In other words, an accelerating body is increasing its velocity so many feet per second every second. Speed is a scalar quantity and does not include direction, but where direction is of no consequence, an acceleration can be said to be a rate of change of speed. As a runner leaves the starting blocks and accelerates from zero to some steady velocity, the time required to reach that velocity is the key to the acceleration. It is likely that this acceleration is not uniform, but it will be assumed to be so for purposes of posing problems.

The student should note the relationship between the first and second laws. It must be emphasized that there cannot be an acceleration without some unbalanced force acting. As soon as the force application ends, the acceleration due to that force also ends and some constant speed would be maintained were it not for the retarding effect of friction or air resistance. Such retarding forces cause a negative acceleration sometimes termed *deceleration.*

The second law is fully expressed by the equation

$$a = \frac{F}{m} \tag{4-1}$$

where F = unbalanced force, lb or N

m = mass of body, slugs or kg

a = uniform or average horizontal acceleration, ft/s^2 or m/s^2

This fundamental formula will be used a number of times in this and later chapters, and so it must be well understood. Suppose that we wish to determine the acceleration resulting when a 20-lb unbalanced force is applied horizontally to a 64-lb object. The object's mass is found by dividing its weight by the gravitational constant:

$$\frac{64 \text{ lb}}{32 \text{ ft/s}^2} = 2 \text{ slugs}$$

Then

$$a = \frac{F}{m} = \frac{20 \text{ lb}}{2 \text{ slugs}} = 10 \text{ ft/s}^2$$

Similarly, if a 20-N force is applied to a 5-kg mass, the resulting acceleration is

$$a = \frac{20 \text{ N}}{5 \text{ kg}} = 4 \text{ m/s}^2 \qquad (1 \text{ N} = 1\text{kg}\cdot\text{m/s}^2)$$

Law 3: The Law of Reaction

For every linear action, there is an equal and opposite reaction.

All forces act in pairs, and one could say that for every force exerted by one body on a second, there is an equal and opposite force reaction by the second body on the first. Two forces must interact; for example, when a sprinter drives against the starting blocks, the blocks push back against his or her feet. Carrying this further, the blocks must in turn push against the ground, which then instantly applies an opposite force against the blocks. If the blocks were to slip, there would be inadequate reaction and the runner would have a poor start.

Persons standing erect exert a force on the ground equal to their weight, and the ground's upward reaction is against the feet. At each succeeding higher joint, there is a reaction to the amount of weight being supported. Obviously, the higher the joint, the less the weight that is borne, as evidenced by the structure of the spinal column, where the lower lumbar vertebrae are much larger than either the higher thoracic vertebrae or the highest, smallest cervical vertebrae. The handstand position alters the normal weight-bearing patterns so that the joints must bear unaccustomed weight. In spacecraft, astronauts must adapt to the strange absence of the normal pressures on the joints.

There are seemingly endless examples of the effects of the third law, and one or two more examples should serve the present purpose. Anyone who has ever fired a shotgun or a military rifle knows about the kick felt in the shoulder as the weapon reacts to the shot or bullet leaving the muzzle. A beginner on roller or ice-skates generally keeps her feet parallel during her efforts to propel herself, and the result is that not much progress is made, on account of the lack of frictional reaction. With learning, the skate is turned at an angle during the push. Without better laboratory equipment, the third law can be tested by standing on an ordinary scale. A vigorous abduction of the arms will momentarily cause the scale to read considerably more than the person's weight. Similarly, from a stance with the arms held overhead, a quick lowering of the arms will show a momentarily lighter weight.

The single skill of heading a soccer ball, shown in Figure 4–2, will serve to illustrate all three laws of motion. As the ball flies through the air, it has an inertia proportional to its mass and would continue eternally in a straight line at a constant speed if it were not acted upon by air resistance and gravity, two forces that negatively accelerate the ball by slowing its speed and changing its direction. The ball is then given an acceleration by the player's head, which applies a force to the ball to change its direction and possibly its speed. The player feels an equal force acting on his head when the ball is met.

Figure 4-2 Heading a soccer ball illustrates Newton's third law of motion. The force applied by the player's head to the ball is also applied by the ball to his head. *Photograph courtesy of the Ohio State University Athletic Publicity Department.*

The ball's effect is a function of its speed as well as its mass. When a ball is wet, its mass is increased and it will have a slightly higher momentum even if the speed is held constant.

VELOCITY AND SPEED

Velocity is defined as displacement per unit of time, and having both magnitude and direction, it is a vector quantity. Speed is expressed as a magnitude only and is therefore a scalar quantity. However, the distinction between speed and velocity is relatively unimportant in most sports situations, and the two terms are often used interchangeably. To make the relationship more clear, consider a girl who ice-skates once around the perimeter of a rink and finishes at her starting point. Since displacement is the straight-line distance between an originating point and a terminating point, the displacement in this case is zero, and therefore the girl's velocity for this trip around the rink is also zero. Here we are obviously interested in her speed and distance and not in her displacement. Had she skated directly across the rink, she would have had a measurable displacement and velocity. One of the few sports in which a form of displacement is more important than distance is football, where a back may run laterally several yards to elude tacklers but actually advances the ball only a yard or two. Here only forward motion is important. Figure 4–3 illustrates the difference between displacement and distance from one point to another.

The velocity symbol v will be used in the motion problems even when speed is meant because there is no commonly used symbol for speed. Following are four meanings of velocity or speed as they will be used in the various motion equations. The symbols are by no means standardized.

Average Velocity

Average velocity ($\bar{v}$) is determined by dividing the total distance covered or the displacement by the total time of the trip. In a mile run of four minutes flat, the runner's average velocity is 22 ft/s, but of course it is unlikely that at every given moment during the race the runner could be timed at exactly that velocity. Not only will each lap probably be run in different times, but for strategic reasons the general pace may be slowed or speeded up several times in any single lap. In fact a beginning marathon runner might very well walk part of the distance and stop frequently for water, and his or her average speed will still be calculated as total distance divided by total time.

Figure 4–3 Distance is represented by the dotted line, traveling a longer route from A to B than is found by measuring the displacement, which is the shortest distance from A to B.

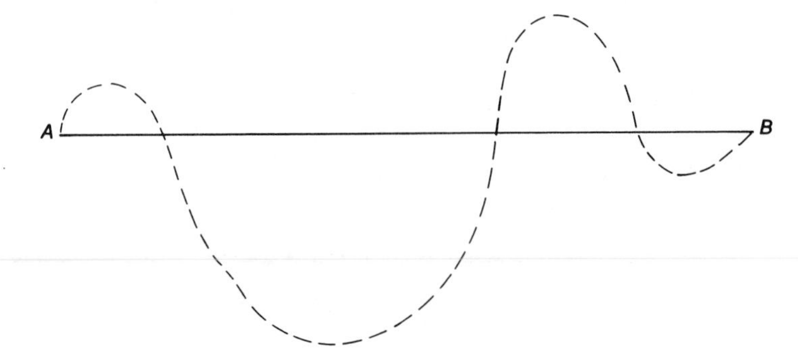

Original Velocity

Original velocity (v_o) is the velocity at the instant that time measurement is begun. Very often, measurement is made of a person or object at rest when the timing process begins, in which case the original velocity is zero.

Final Velocity

Final velocity (v_f) is the velocity that is measured at the end of some distance or some period of time. In a race, this may be a split time or the time at the finish line.

Instantaneous Velocity

Instantaneous velocity (v) is the velocity measured at any particular time or place of a trip or race. A speedometer shows the instantaneous speed of a car at any given moment. For runners or swimmers, the instantaneous velocity might be determined as the distance covered in a very short period of time during any segment of a race. What is necessary is that the time interval be so short that there is nearly no chance for the velocity to change appreciably during the measurement. Of course, the interval must be measurable with available clocks. It is easier to position timers between two close points than it is to take two times and try to measure the distance covered. The calculation of instantaneous velocity is made considerably more simple with cinematography.

EQUATIONS FOR MOTION BEGINNING FROM REST

When bodies begin moving from a resting state and accelerate uniformly, or when average acceleration ($\bar{a}$) is known, the following equations can be used to solve for

v_f Final velocity or speed.

$\bar{v}$ Average velocity or speed.

a Uniform or average acceleration.

t Time.

d Distance. (If it is understood that the motion is in a straight line only, then *displacement* is an acceptable term for the symbol d.)

Final Velocity Starting from Rest

$$v_f = at \quad (4\text{–}2)$$

gives the velocity after any particular time during a period of uniform acceleration.

$$v_f^2 = 2ad \quad (4\text{–}3)$$

gives the velocity after some distance is covered during a period of uniform acceleration.

Average Velocity

$$\bar{v} = \frac{v_f}{2} \quad (4\text{–}4)$$

gives the average velocity up to the moment that the final velocity is measured following a period of uniform acceleration.

$$\bar{v} = \frac{d}{t} \quad (4\text{–}5)$$

gives the average velocity when the nature of the acceleration is *not* a consideration. There may have been unaccelerated periods during the trip.

Acceleration

$$a = \frac{v_f}{t} \quad (4\text{–}6)$$

gives the uniform or the average acceleration when the velocity at any particular time is known. See Eq. (4–2). Since acceleration is the rate of change of velocity, it is important here to remember that in this equation $v_o = 0$.

$$a = \frac{v_f^2}{2d} \quad (4\text{–}7)$$

gives the uniform or the average acceleration when the velocity after a particular distance is known. See Eq. (4–3).

Time

$$t = \frac{v_f}{a} \qquad \textbf{(4–8)}$$

gives the time when the final velocity and the rate of uniform or average acceleration are known. See Eq. (4–2).

$$t = \frac{d}{\bar{v}} \qquad \textbf{(4–9)}$$

gives the time when the distance covered and the average velocity are known. See Eq. (4–5).

Distance (or Displacement in a Straight Line)

$$d = \bar{v}t \qquad \textbf{(4–10)}$$

gives the distance traveled when the average velocity for a period of time is known. See Eq. (4–5).

$$d = \frac{v_f^2}{2a} \qquad \textbf{(4–11)}$$

gives the distance when the final velocity and the rate of uniform or average acceleration are known. See Eq. (4–3).

$$d = \tfrac{1}{2}at^2 \qquad \textbf{(4–12)}$$

gives the distance covered during a particular time of uniform acceleration. It is derived by inserting the value of $\bar{v}$ from Eq. (4–4) into Eq. (4–10) and then substituting the value of v_f from Eq. (4–2). This equation can be rearranged to determine acceleration and time as follows:

$$t^2 = \frac{2d}{a} \qquad \textbf{(4–13)}$$

$$a = \frac{2d}{t^2} \qquad \textbf{(4–14)}$$

EQUATIONS FOR BODIES WITH SOME PREVIOUS MOTION

In those cases where a body is already in motion at the beginning of measurement, the original velocity v_o must be considered.

$$v_f = v_o + at \qquad (4\text{–}15)$$

Compare to Eq. (4–2).

$$v_f^2 = v_o^2 + 2ad \qquad (4\text{–}16)$$

Compare to Eq. (4–3).

$$\bar{v} = \frac{v_o + v_f}{2} \qquad (4\text{–}17)$$

Compare to Eq. (4–4).

$$a = \frac{v_f - v_o}{t_f - t_o} \qquad (4\text{–}18)$$

Compare to Eq. (4–6).

$$d = v_o t + \tfrac{1}{2} at^2 \qquad (4\text{–}19)$$

Compare to Eq. (4–12).

The student need not be apprehensive about the number of equations to be learned, as they are all based upon the first four, Eqs. (4–2) through (4–5), and only those need be memorized. The formula to be used obviously depends upon what is sought and what information is given. The uses of these equations may be made clearer through a few example problems.

Example 4–1. A sprinter accelerates uniformly from 0 to 20 ft/s in 3 s.

1. What is the average acceleration?
2. What is the average velocity?
3. How far does the sprinter run in this time?

1. Use Eq. 4–6:

$$a = \frac{v_f}{t}$$

$$= \frac{20 \text{ ft/s}}{3 \text{ s}} = 6.66 \text{ ft/s}^2$$

2. Use Eq. 4–4:

$$\bar{v} = \frac{v_f}{2}$$

$$= \frac{20 \text{ ft/s}}{2} = 10 \text{ ft/s}$$

3. Use Eq. 4–10:

$$d = \bar{v}t$$

$$= 10 \text{ ft/s} \times 3 \text{ s} = 30 \text{ ft}$$

Example 4–2. A cyclist who is moving 30 km/h puts on the brakes and uniformly decelerates to a stop in 3 s. What is the deceleration (negative acceleration)?

Use Eq. 4–18:

$$a = \frac{v_f - v_o}{t}$$

$$= \frac{0 - 30 \text{ km/h}}{3 \text{ s}} = -10 \text{ km/h/s}$$

Example 4–3. What is the velocity of a sprinter who is uniformly accelerating from rest at 3m/s² for 1.5 s?

Use Eq. 4–2:

$$v_f = at$$

$$= 3 \text{ m/s}^2 \times 1.5 \text{ s} = 4.5 \text{ m/s}$$

Example 4–4. What will be the distance covered by a sprinter who accelerates from rest at a uniform rate of 9 ft/s² until reaching a velocity of 20 ft/s? What else can be determined from these data?

Use Eq. 4–11:

$$d = \frac{v_f^2}{2a}$$

$$= \frac{(20 \text{ ft/s})^2}{2 \times 9 \text{ ft/s}^2}$$

$$= \frac{400 \text{ ft}^2/\text{s}^2}{18 \text{ ft/s}^2} = 22.2 \text{ ft}$$

Since we know the velocity and the acceleration, time may be found with Eq. (4–8):

$$t = \frac{v_f}{a}$$

$$= \frac{20 \text{ ft/s}}{9 \text{ ft/s}^2} = 2.22 \text{ s}$$

The average velocity from rest to a given velocity is found with Eq. (4–4):

$$\bar{v} = \frac{v_f}{2}$$

$$= \frac{20 \text{ ft/s}}{2} = 10 \text{ ft/s}$$

Example 4–5. A woman is moving along on a motorcycle at 15 mi/h and then accelerates uniformly to 40 mi/h in an 11-s time period.

1. What is her acceleration?
2. How far does she travel in those eleven seconds?

1. Use Eq. (4–18):

$$a = \frac{v_f - v_o}{t_f - t_o}$$

$$= \frac{40 \text{ mi/h} - 15 \text{ mi/h}}{11 \text{ s}} = \frac{25 \text{ mi/h}}{11 \text{ s}}$$

$$= (2.27 \text{ mi/h})/\text{s}$$

which may be converted to feet per second per second by multiplying by 1.467, which gives 3.33 ft/s².

2. Use Eq. (4–19):

$$d = v_o t + \tfrac{1}{2} a t^2$$

but first convert the original velocity of 15 mi/h into feet per second by multiplying by 1.467:

$$d = 22 \text{ ft/s} \times 11 \text{ s} + \frac{3.33 \text{ ft/s}^2 \times (11 \text{ s})^2}{2}$$

$$= 242 \text{ ft} + 201.5 \text{ ft}$$

$$= 443.5 \text{ ft}$$

FREE-FALLING BODIES

Until Galileo postulated otherwise, scientists through the sixteenth century believed the Aristotelian concept that heavy objects fall at a faster rate than light objects. The invention of the air pump about 1650 provided the means to substantiate Galileo's theory. When air was pumped out of a container to create a vacuum, it was demonstrated that a coin and a feather fell at the same speed.

The study of linear motion includes the behavior of bodies falling freely in space. Except in a few cases such as sky diving and ski jumping, the effects of air resistance are usually ignored and, for ease of calculation, the round-figure gravitational acceleration of 32 ft/s^2 or of 9.8 m/s^2 is commonly used. This is the acceleration that affects all falling or thrown objects, as well as the projected human body, and it should not be necessary to do more than call attention to the fact that if a body falls a good deal farther than is usual in sports, a terminal velocity may be reached where the air resistance equals the weight of the falling body, at which point there is no unbalanced force and thus no further acceleration. This terminal speed can be reached by a skydiver, but the chutist can affect when it occurs by using body manipulations to increase or decrease air resistance.

The force of gravity acts identically on both ascending and descending bodies regardless of their weight and always acts in a vertically downward direction. When a ball is thrown straight up into the air, it loses velocity (negatively accelerates) at the rate of 32 ft/s^2 (9.8 m/s^2) until it reaches a peak height where its vertical velocity is momentarily zero, whereupon it immediately begins to fall and gain velocity (positively accelerate) at the same 32 ft/s^2 (9.8 m/s^2) rate. By the time the ball falls back to the level from which it was thrown, its velocity will have reached that which it had at the instant of release, and the time needed for it to rise to peak height will be the same as the time taken in descent.

By definition, acceleration is the change of velocity per unit of time, and so the term can be used in reference to bodies which are decreasing in speed as well as to those whose speed is increasing. However, for the sake of clarity, it may be desirable to say that bodies which are losing speed are either *decelerating* or *negatively accelerating* and to reserve the term *accelerating* for bodies which are increasing in speed.

All the equations previously given for uniformly accelerating bodies can also be used for falling bodies, but some of the symbols are normally changed so as to make free-falling situations somewhat distinct and to recognize the uniformity of gravitational acceleration. While there does not appear to be any universally consistent symbol usage, in this text it will be the practice

To use s in place of d when discussing the height of a rising or falling body

To use g in place of a when discussing gravitational acceleration

To use v_y in place of v_f when discussing the final vertical velocity of a falling or rising body or the vertical velocity component when the body has some horizontal speed as well.

Equations for Bodies Falling or Rising from Rest

$$v_y = gt \tag{4-20}$$

gives the vertical velocity after any particular time of falling from rest and the *initial* upward velocity for bodies that rise for a specified time before beginning to fall. See Eq. (4–2). When time to rise or to fall is known, the vertical velocity is obtained as the product of the time and 32 ft/s^2 or 9.8 m/s^2.

$$v_y^2 = 2gs \tag{4-21}$$

gives the vertical velocity at a particular height for a descending body and gives the *initial* upward velocity needed for a body to rise to a specified height from rest. See Eq. (4–3). Take the square root of the product of the given height and 64 in the British system or 19.6 in the MKS system.

$$s = \frac{v_y^2}{2g} \tag{4-22}$$

gives the height attained when the initial vertical rising or falling velocity at the same instant is known. See Eqs. (4–11) and (4–21).

$$s = \tfrac{1}{2} gt^2 \tag{4-23}$$

gives the vertical distance fallen in a particular time, usually stated in seconds. This equation may be used to find peak height when the time to rise is known. See Eq. (4–12).

$$s = \tfrac{1}{2}g\,(2t - 1) \tag{4–24}$$

gives the distance fallen between one instant of time and the preceding second. To find the distance fallen during a given second of its descent, multiply the distance a body falls in the first second, that is, 16 ft, by 1 less than twice the total number of seconds of descent that have elapsed.

$$t = \frac{v_y}{g} \tag{4–25}$$

gives the time it takes for a falling body to achieve a particular vertical velocity or the time a body will rise when the initial velocity is known. See Eq. (4–20).

$$t^2 = \frac{2s}{g} \tag{4–26}$$

gives the square of the time it takes to rise to a particular peak height or to fall from rest a specified distance. See Eq. (4–23).

Table 4–1 illustrates the relationships between falling distance, time, and vertical velocity for objects falling from rest. It can be seen that, if the effects of air resistance are ignored, falling speed is directly proportional to the lapse of time; that is, after 1s, $v_y = g$; after 2s, $v_y = 2g$; and so on Eq. (4–20)). The distance fallen is proportional to the square of the time (Eq. 4–23)).

TABLE 4-1

Relationships between Distance, Time, and Velocity for Objects Falling from Rest Based on a *g* Value of 32 ft/s² or 9.8 m/s²

s	t	v_y
16 ft (4.9 m)	1 s	32 ft/s (9.8 m/s)
64 ft (19.6 m)	2 s	64 ft/s (19.6 m/s)
144 ft (44.1 m)	3 s	96 ft/s (29.4 m/s)
256 ft (78.4 m)	4 s	128 ft/s (39.2 m/s)

Because any examples that can be given for rising and falling bodies are very similar to those given for linear motion equations, only two will be used.

Example 4–6. If air resistance is not taken into account, how far will any object fall from rest in 5 s? What will be its velocity at that point?

Use Eq. (4–23):

$$s = \tfrac{1}{2}gt^2$$

$$= 16\ \text{ft/s}^2 \times 25\ \text{s}^2 \quad \text{or} \quad 4.9\ \text{m/s}^2 \times 25\ \text{s}^2$$

$$= 400\ \text{ft} \quad \text{or} \quad 122.5\ \text{m}$$

Use Eq. (4–20):

$$v_y = gt$$

$$= 32\ \text{ft/s}^2 \times 5\ \text{s} \quad \text{or} \quad 9.8\ \text{m/s}^2 \times 5\ \text{s}$$

$$= 160\ \text{ft/s} \quad \text{or} \quad 49\ \text{m/s}$$

Alternatively, Eq. (4–21) could be used to find v_y.

Example 4–7. How many meters will any object fall from rest in 3 s?

Use Eq. (4–23):

$$s = \tfrac{1}{2}gt^2 = \tfrac{1}{2} \times 9.8\ \text{m/s}^2 \times (3\ \text{s})^2$$

$$= 44.1\ \text{m}$$

PROJECTILES

A projectile is any body that is impelled by some force and then continues to move through the air by its own inertia. A thrown ball, an airborne long-jumper, and a released arrow are all examples of projectiles, and what they have in common with all projectiles is the gravitational force that acts identically on each.

Because of the effect of gravity, a projectile will follow a parabolic course unless the projection is exactly vertical. This path may be altered to varying degrees by air resistance, the amount of which depends upon the object's size, weight, shape, surface characteristics, and speed. The effects of air resistance and of any airfoil qualities of the object will be ignored for now.

The study of projectiles requires the application of principles related to both linear motion and free-falling bodies. All sports projectiles, including human ones, will be considered here to be particles in flight, because the path of a single point, represented by a mass center, is easier to plot.

The gravitational force that acts on a projectile is *entirely* independent of the object's horizontal speed. When an object is projected at some angle relative to the horizontal, there are two independently operating velocity components acting perpendicularly to one another. The vertical velocity vector v_y is affected by gravity, and therefore, its value will vary from maximum at the instant of release of the object to zero at the peak of the projectile's flight. The v_y at release determines how high an object will rise, and it should be noted that the time to rise to a peak equals the time to fall back to the level of release.

On the other hand, the horizontal velocity component v_x is constant in value in accordance with Newton's first law and is actually nothing more than average velocity $\bar{v}$. Once a projectile's horizontal velocity is known, it can be assumed that it will remain the same throughout its flight, since we have already stated that the retarding effects of air resistance will be overlooked.

The resultant of a plotting of the v_x and v_y vectors is v_θ, the object's actual velocity in the direction of projection as caused by an unbalanced force applied up to the instant of release. This resultant v_θ can be determined trigonometrically or graphically.

To illustrate the independence of the component velocities, look at Figure 4–4, which shows a bullet fired horizontally at the same time that another bullet is simply dropped from the height of the gun muzzle. Because gravity acts equally on both bullets regardless of their horizontal velocities, both bullets will hit the ground at the same moment. Both have the same v_y and the time in the air will be the same for both. This is the reason why we

Figure 4–4 A bullet fired horizontally and a bullet dropped at the same time from the height of the muzzle will strike the ground at the same instant.

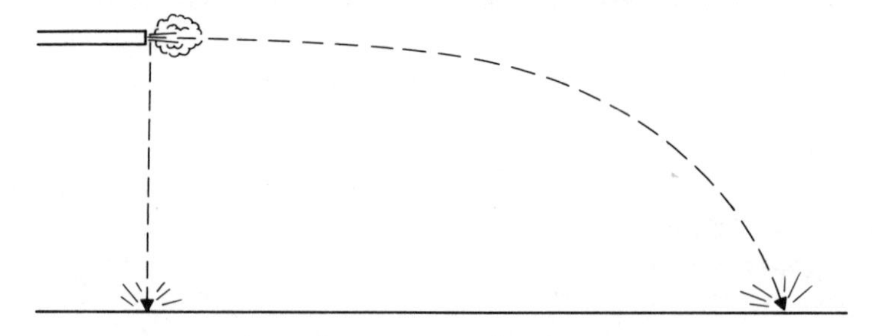

must seek elevation when projectile distance is an objective. Elevation gains time in the air so that the horizontal velocity can be most effective.

Where achievement of maximum distance is the prime consideration, as it is in shot-putting, a projection angle between 40° and 45° is theoretically ideal, because within this range of angles the vertical and horizontal components of velocity are nearly the same and provide the best combination of time in the air and forward velocity. An object projected at any angle below 45° will have a higher horizontal velocity than it would have at an angle exceeding 45°. But the time in the air, which is a function of the vertical velocity, will be less in the first instance.

When projected at each of two complementary angles at the same velocity, an object's horizontal displacement will be the same if air resistance is not considered. (The effect of the height at which the projectile is released will be considered shortly.) For example, an object projected at 30° should travel the same distance as one released at 60° with the same speed. In the first case, the horizontal velocity is high but time in the air is short; in the latter case, there is a great deal more time in the air, but the horizontal velocity is less. Which angle is preferred may be dictated by game strategy, as when a tennis player hits a deep defensive lob to the opponent's base line in order to gain time to recover to play the return.

In only a few sports is it necessary for a person or an object to achieve maximum distance, height, or velocity. Most often a throw will be less than maximal and emphasis will be on accuracy. This is the case in basketball, baseball, darts, horseshoes, and football, to name but a few sports. The appropriate angle of release depends upon the sport, and within each sport it may also depend upon the game situation or the individual preference of the performer. The human being cannot generate equal amounts of force at every angle because of the added work that must be done as the angle is elevated above the horizontal. Even where distance is desired, Miller and Nelson state that a high release velocity is the most important single factor.[1]

As was stated earlier, the parabolic path can be changed by the airfoil characteristics of the projectile or by air resistance. Light objects that are in the air several seconds will be affected much more than will heavy objects in the air a few seconds. Many a baseball hit high and deep has gone foul because of a wind. A 16-lb shot is not easily moved off course by a breeze. A badminton shuttlecock, even when driven hard, will drop rather sharply as the air resistance takes effect.

It is both interesting and important to note that once a person jumps or dives, the parabolic path of his or her center of gravity is set, and no movements of the torso or limbs while in the air will change that path. Only

[1] Doris I. Miller and R. C. Nelson, *Biomechanics of Sport* (Philadelphia: Lea & Febiger, 1973), p. 80.

external unbalanced forces can alter the direction of the body's mass center. The athlete can, of course, rotate various body parts around the body's center of gravity to accomplish a certain stunt or to put the body into a particular position for landing, as in the long jump. The underlying principle for this will be discussed in Chapter 5.

The typical path of a projectile in the absence of air resistance is illustrated in Figure 4–5, and for simplicity, the projection is depicted as two-dimensional.

Basically, all projectile problems seek to find one or more of the following:

v_θ	Projecting velocity
v_x	Horizontal velocity component
v_y	Vertical velocity component
θ	Projection angle relative to the horizontal
R	Horizontal range (displacement)
t	Time to rise to a peak or to fall from a peak
T	Total airborne time
s	Peak height achieved

Because the component velocities v_x and v_y always act at right angles to one another, as shown in Figure 4–5, trigonometry can be used to solve projectile problems. The side adjacent to angle θ is represented by v_x; v_y represents the opposite side; and v_θ is the hypotenuse of the right triangle.

Figure 4–5 The parabolic path of a projectile with superimposed velocity vectors.

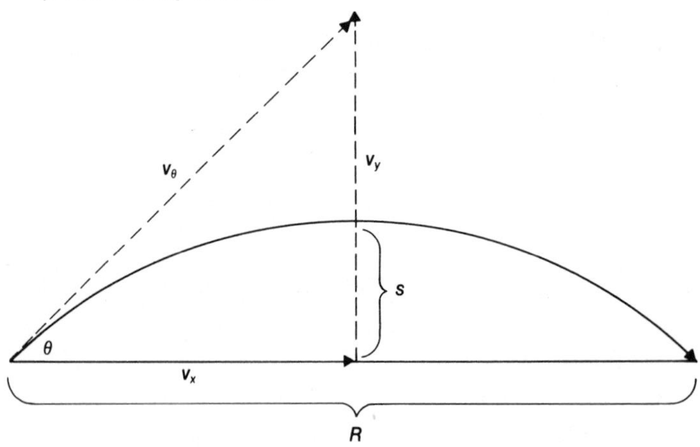

v_x is calculated by multiplying the projecting velocity by the cosine of the angle formed between the projection vector and the horizontal. Thus, $v_x = v_\theta \cos\theta$, since cosine = (adjacent side)/hypotenuse.

v_y is obtained by multiplying the projecting velocity by the sine of the angle. Thus, $v_y = v_\theta \sin\theta$, since sine = (opposite side)/hypotenuse.

v_θ is determined by either of the two above processes if the angle and one component are known, or by use of the Pythagorean theorem if both components are known. Thus, $v_\theta^2 = v_x^2 + v_y^2$.

Peak height reached at the top of a parabola is a function of the vertical velocity component and is not affected by the projectile's horizontal speed. This height is found with Eq. (4–22) or (4–23), the former if the vertical velocity is given and the latter if the time is known. Note that in Eq. (4–23), t represents either the time to rise to the peak or the time to fall back. Total time in the air cannot be used as the figure for t in the equation $s = \frac{1}{2}gt^2$.

The time to rise or to fall is found from either Eq. (4–25) or Eq. (4–26), the former if the vertical velocity is known and the latter if the height is given. Total time is found by doubling t or by using

$$T = \frac{2v_y}{g} \tag{4–27}$$

Range or horizontal displacement of a projectile can be determined by any of the following identical methods. The available data will indicate the preferred equation.

$$R = v_x T \tag{4–28}$$

gives the horizontal displacement of a projectile when the total time in the air and the horizontal velocity component are known. Compare to Eq. (4–10); v_x here can be defined as the average velocity along the line of displacement. This equation may be written $R = v_\theta \cos\theta\ T$.

$$R = \frac{2v_x v_y}{g} \tag{4–29}$$

gives the horizontal displacement when the two component velocities are known. It is apparent that this equation is the same as Eq. (4–28) when the value of T from Eq. (4–27) is considered. This equation may be written

$$R = \frac{2(v_\theta \cos\theta)(v_\theta \sin\theta)}{g}$$

$$R = \frac{v_\theta^2 \sin 2\theta}{g} \tag{4–30}$$

gives the horizontal displacement when only the projecting velocity and angle of release are known. It is necessary to first double the angle θ before looking up its sine (or to look up the sine of double the complementary angle of θ). Note that sin 2θ equals 2 sin θ cos θ, and so the equation could be written

$$R = \frac{2v_\theta^2 \sin\theta \cos\theta}{g}$$

which is identical to the alternate way of writing Eq. (4–29).

Most projectile problems have several common elements, and the ability to solve one such problem is a pretty fair indication that most others can be handled with some ease.

Example 4–8. An archer shoots an arrow at an angle of 27° to the horizontal and scores a bull's-eye on a target whose center is on the same level as that of the release from the bow. The arrow has a release velocity of 96 ft/s.

1. How long will the arrow be in the air?
2. What will be its greatest height in flight?
3. What will be the arrow's displacement?

If these questions are to be answered in the order in which they are asked, start by determining the vertical velocity component of the arrow using $v_y = v_\theta \sin\theta$:

$$96 \text{ ft/s} \times .454 = 43.58 \text{ ft/s}$$

Then use Eq. (4–27) to find the total time in the air:

$$T = \frac{2v_y}{g}$$

$$= \frac{2 \times 43.58 \text{ ft/s}}{32 \text{ ft/s}^2} = 2.72 \text{ s}$$

There are two ways to find height, Eq. (4–22) or Eq. (4–23):

$$s = \frac{v_y^2}{2g} = \frac{(43.58 \text{ ft/s})^2}{64 \text{ ft/s}^2} = \frac{1899.22 \text{ ft}}{64} = 29.68 \text{ ft}$$

This is the greatest height above the point of release. The second method is:

$$s = \tfrac{1}{2}gt^2$$

$$= 16 \text{ ft/s}^2 \times (1.36 \text{ s})^2 = 29.6 \text{ ft}$$

The range may be found by any of three methods:

1. Use $R = v_x T$. T is already known, and so v_x must be found.

$$v_x = v_\theta \cos\theta$$
$$= 96 \text{ ft/s} \times .891 = 85.54 \text{ ft/s}$$

Now use v_x and T to find R:

$$R = 85.54 \text{ ft/s} \times 2.72 \text{ s}$$
$$= 233 \text{ ft}$$

2. Use Eq (4–29) after finding v_x and v_y by methods already described.

$$R = \frac{2v_x v_y}{g}$$
$$= \frac{(2)(85.54 \text{ ft/s})(43.58 \text{ ft/s})}{32 \text{ ft/s}^2} = 233 \text{ ft}$$

3. Eq. (4–30) should be the equation of choice if displacement is the only information sought, because all the needed data are provided in the original statement of the problem. Two times the angle is 54°, for which the sine is .809.

$$R = \frac{v_\theta^2 \sin 2\theta}{g} = \frac{(96 \text{ ft/s})^2(.809)}{32 \text{ ft/s}^2}$$
$$= \frac{9216 \text{ ft}^2/\text{s}^2 \times .809}{32 \text{ ft/s}^2}$$
$$= 233 \text{ ft}$$

The foregoing equations are of no value in dealing with sports objects whose flight paths are easily affected by air resistance. In the cases of footballs thrown deep, badminton shuttlecocks, boomerangs, and discuses, to name a few, it would be very difficult to predict flight paths from such data as projecting velocity, height attained, or horizontal displacement.

Now that the basic projectile principles have been presented, it should be pointed out that in sports, very often the projected object will land at a level lower than that from which it is launched. For example, a shot may leave the putter's hand 7 ft above ground level and then land at ground level, where its displacement is measured (Figure 4–6). The equations given

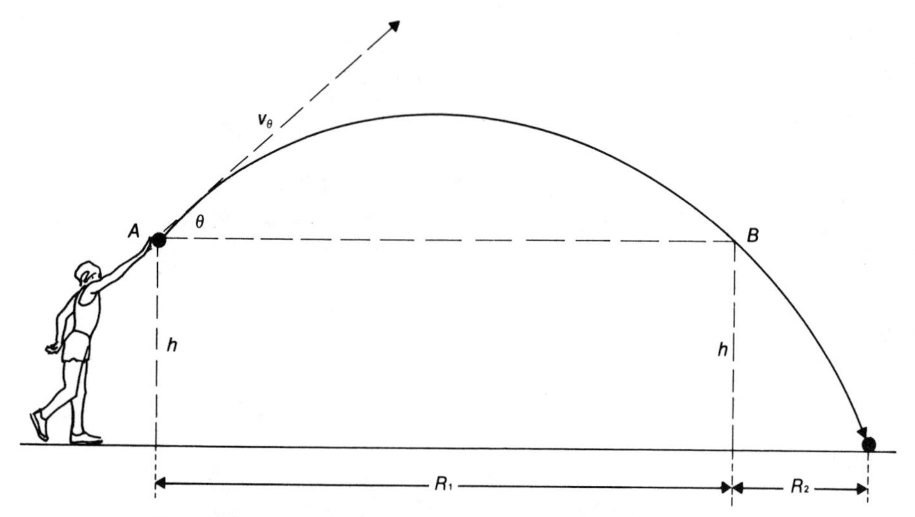

Figure 4-6 A projectile released from a level higher than that at which it will land.

previously cannot all be used to compute the range of the shot, because of the added displacement, which has been labeled R_2.

For those projectile problems in which the impact point is lower than the point of release, the following equation must be used:

$$R_1 + R_2 = \frac{v_x v_y + v_x \sqrt{v_y^2 + 2gh}}{g} \qquad \textbf{(4–31)}$$

where R_1 = displacement from release height at point A in Figure 4–6 to point B, which is the same height above ground

R_2 = additional displacement between point B and impact point C on the ground

h = height from landing level to level of release

v_x and v_y = horizontal and vertical velocity components of projecting velocity

There are other equations that yield the same results, but only one alternative, already known method will be suggested, Eq. (4–28). At first glance it might appear that $R = v_x T$ is too simple compared with the complex-looking Eq. (4–31), but the eight steps that need to be computed will show that there is little to be gained by using it:

1. Find v_x as the product of v_θ and cos θ.
2. Find v_y as the product of v_θ and sin θ.
3. Find the time to rise with Eq. (4–25), $t = v_y/g$.
4. Find peak height from release level with Eq. (4–23), $s = \frac{1}{2}gt^2$.
5. Find the total height from ground level by adding h and s.
6. Find the time required for an object to fall the distance calculated in step 5 by using Eq. (4–26), $t^2 = 2s/g$.
7. Add the times found in steps 3 and 6 to get the total time.
8. The $R_1 + R_2$ displacement is then the product of the v_x from step 1 and the T from step 7.

One worked problem might help clarify how the two methods can be utilized in solving a projectile range when the release height is above the landing level.

Example 4–9. Consider a softball thrown for distance which leaves the thrower's hand at a velocity of 60 ft/s at an angle of 43°. At the instant of release, the hand is 6 ft above ground level. Using two methods, find the total distance of the throw as measured from a point on the ground directly below the release to the place of landing. Reference to Figure 4–6 might be helpful.

1. Use Eq. (4–31). The only unknown values are the two velocity components. Following previous procedures,

$$v_y = v_\theta \sin\theta$$

$$= 60 \text{ ft/s} \times .682 = 40.9 \text{ ft/s}$$

$$v_x = v_\theta \cos\theta$$

$$= 60 \text{ ft/s} \times .731 = 43.9 \text{ ft/s}$$

$$R_1 + R_2 = \frac{(43.9)(40.9) + 43.9\sqrt{(40.9)^2 + (2)(32)(6)}}{32} \text{ ft}$$

(Since it is evident from Eq. (4–31) that all units cancel out to yield a result in units of displacement, the internal units can be left out here because they are unnecessary, so long as it is shown that the result will be in feet.)

$$R_1 + R_2 = \frac{1795.5 + 43.9\sqrt{2056.81}}{32}\ \text{ft}$$

$$= \frac{1795.5 + (43.9 \times 45.35)}{32}\ \text{ft}$$

$$= \frac{1795.5 + 1991}{32}\ \text{ft}$$

$$= 118.3\ \text{ft}$$

2. Use Eq. (4–28). Follow the eight-step process outlined previously. Since we found v_x and v_y above, start with step 3 and find time to rise with Eq. (4–25):

$$t = \frac{v_y}{g} = \frac{40.9\ \text{ft/s}}{32\ \text{ft/s}^2} = 1.28\ \text{s}$$

Find peak height from release with Eq. (4–23):

$$s = \tfrac{1}{2}gt^2 = (\tfrac{1}{2})(32\ \text{ft/s}^2)(1.28\ \text{s})^2 = 26.2\ \text{ft}$$

Total height = (26.2 + 6) ft = 32.2 ft. Find the time needed to fall 32.2 ft with Eq. (4–26):

$$t^2 = \frac{2s}{g} = \frac{64.4\ \text{ft}}{32\ \text{ft/s}^2} = 2.0125\ \text{s}^2$$

$$t = \sqrt{2.0125} = 1.42\ \text{s}$$

Total airborne time = 1.42 s + 1.28 s.

$$T = 2.7\ \text{s}$$

Displacement of the softball is:

$$R = v_xT = 43.9\ \text{ft/s} \times 2.7\ \text{s}$$

$$= 118.5\ \text{ft}$$

Such problems as this once took considerable time to solve, but inexpensive pocket calculators are now so commonly available that even a long-drawn-out procedure can be handled with ease. (A calculator also saves the trouble of rounding off the results of each step and thus can prevent the kind of discrepancy that exists between the two final results in Example 4–9.)

Trigonometric function tables as presented in Appendix C are probably already unnecessary and will go the way of the slide rule.

When determining the impact velocity of an object thrown horizontally from some height, one cannot use v_y as the velocity of impact, because v_x is also a contributor. Thus, a ball thrown horizontally at 50 ft/s from a height of 100 ft will strike the ground at a higher velocity than a ball that is simply dropped from that height, even though the falling time for both is the same. It is necessary to use the Pythagorean theorem to calculate the impact velocity v_θ, and it is interesting to note that the path of the ball is just one half of a parabola. Thus the impact velocity is the same as would be the projecting velocity under the same conditions of s = 100 ft and v_x = 50 ft/s.

Determine v_y with Eq. (4–21):

$$v_y^2 = 2gs = (2)(32 \text{ ft/s}^2)(100 \text{ ft}) = 6400(\text{ft/s})^2$$

$$v_y = \sqrt{6400 \ (\text{ft/s})^2} = 80 \text{ ft/s}$$

Knowing v_x and v_y, you can find v_θ by

$$v_\theta^2 = v_x^2 + v_y^2$$

$$= (50^2 + 80^2)(\text{ft/s})^2 = 8900 \ (\text{ft/s})^2$$

$$v_\theta = \sqrt{8900 \ (\text{ft/s})^2} = 94.3 \text{ ft/s}$$ compared with 80 ft/s for a ball dropped 100 ft

Can you find the angle at which the ball lands?

MOMENTUM

Linear momentum can be defined as the quantity of motion *or* as the product of a moving body's mass and linear velocity; thus

$$P = mv \tag{4–32}$$

In the British system of measurements, the momentum unit is slug-feet per second in a particular direction. In the SI system the unit is kilogram-meters per second in a particular direction. Obviously, momentum has both magnitude and direction and is, therefore, a vector quantity.

A running football player has momentum, and he can increase it simply by running faster. In the off-season, he might add some weight, and provided that he can still maintain his speed, his momentum will increase. Any such increase will make him harder to stop and will increase his ability to stop an opponent.

The effect that a baseball player's bat has on a pitched ball is partly a function of the bat's angular momentum. The player can increase the momentum of the bat by swinging it faster, by using a heavier bat, or both. Presumably, keeping a firm grip at impact might add some of the batter's own mass to that of the bat.

In sports, the only real importance of momentum is in its role in collision situations, as when two players collide, when a bat meets a ball, and so on. Collisions will be covered in Chapter 6. The topic of momentum is brought up now because of its part in the following discussion of impulse.

IMPULSE

Whenever the momentum of an object is suddenly changed, the change is caused by an *impulse,* which is nothing more than a restatement of Newton's second law. The product of the average force applied to an object and the time interval during which the force acts is the definition of this vector quantity called impulse. Eq. (4–1) thus becomes

$$Ft = m(v_f - v_o) \tag{4-33}$$

The units are pound-seconds (lb•s) or Newton-seconds (N•s).

If it is necessary to determine how much force is required to achieve a particular change in momentum in a given, generally very short, time, Eq. (4–33) is altered by dividing both sides by time to get

$$F = \frac{m(v_f - v_o)}{t} \tag{4-34}$$

which is really $F = ma$ stated in terms of momentum change.

Specifically, the concern here is with the product of the force and the time of contact. The shorter the time interval that a force can be applied to achieve a desired momentum change, the greater the force must be. In other words, a given change in momentum can be the result of a small force applied for a long time or of a large force acting for a short time. Consequently, if an athlete has a limited force capability, as all athletes do, it is necessary to increase the time of application if there is to be an increase in the change of momentum in the object being acted upon.

The time of force application and the potential for moderation of the force magnitude cannot always be manipulated by a performer. In shot-putting, the time of force application was increased by the O'Brien method introduced a number of years ago. By comparison, a baseball bat has a very short contact time with a ball and this time cannot be controlled by the batter. In such a case, impulse can be increased only by strengthening the muscles or by swinging the bat faster.

With present high-speed cinematographic equipment, it is possible to measure the contact time of a force. The magnitude of an impulse can be gauged by the resulting change in momentum that is caused.

Example 4–10. A 1.6-oz golf ball is struck and the contact time between club head and ball is 0.0005 s. If the ball's velocity as it leaves the tee is 200 ft/s,

1. What is the force?
2. What is the impulse?
3. Through what distance are the club and the ball in contact?

1. Use Eq. (4–34):

$$F = \frac{m(v_f - v_o)}{t}$$

Convert 1.6 oz into pounds and then into slugs:

$$\frac{1.6 \text{ oz}}{16 \text{ oz/lb} \times 32 \text{ ft/s}^2} = 0.003125 \text{ slugs}$$

$$F = \frac{0.003125 \text{ slugs} \times 200 \text{ ft/s}}{0.0005 \text{ s}} = 1250 \text{ lb}$$

2. The impulse is easily determined:

$$\text{Impulse} = Ft = 1250 \text{ lb} \times 0.0005 \text{ s}$$

$$= 0.625 \text{ lb} \cdot \text{s}$$

3. To obtain contact distance, find the ball's average velocity during contact by the club using Eq. (4–4) and insert that value into Eq. (4–10).

$$\bar{v} = \frac{v_f}{2} = \frac{200 \text{ ft/s}}{2} = 100 \text{ ft/s}$$

$$d = \bar{v}t = 100 \text{ ft/s} \times 0.0005 \text{ s} = 0.05 \text{ ft}$$

This example stated in SI units would be

$$\text{Ball's mass} = 0.0454 \text{ kg}$$

$$\text{Ball's velocity} = 61 \text{ m/s}$$

$$\text{Time} = 0.0005 \text{ s}$$

Thus,

$$F = \frac{0.0454 \text{ kg} \times 61 \text{ m/s}}{0.0005 \text{ s}} = 5539 \text{ N}$$

$$Ft = 5539\ N \times 0.0005 \text{ s} = 2.77 \text{ N}\bullet\text{s}$$

$$\bar{v} = \frac{61 \text{ m/s}}{2} = 30.5 \text{ m/s}$$

$$d = 0.015 \text{ m}$$

FORCES APPLIED AT AN ANGLE

It is appropriate now to point out that Eqs. (4–1) and (4–34) are valid only for horizontally directed forces and accelerations. When an object is elevated by a force applied at any angle between 0° and 90°, a correction must be added to deal with the effects of gravity. This correction represents the additional force necessary to lift a particular component of the object's weight, the F_y component. Equation (4–1) thus becomes:

$$F = ma + mg \sin \theta \qquad \text{(4–35)}$$

where θ = angle of elevation

mg = object's weight

Similarly, Eq. (4–34) becomes:

$$F = \frac{m(v_f - v_o)}{t} + mg \sin \theta \qquad \text{(4–36)}$$

A car traveling at a constant speed along a level road will lose speed when it starts to go up a hill, and so more engine force must be applied via the gas pedal if speed is to be maintained. To put a shot horizontally takes much less effort than it would to put it at an angle of 40°, but since elevation is needed to achieve distance, the maximum available force must be applied at this angle even though the resulting acceleration will be less than at smaller projection angles.

The correction becomes more important as the mass to be moved is larger; that is, it is not too significant when dealing with a golf ball, but it is quite important when analyzing the shot put. The greater the angle of elevation, the more the force that is required *if* a desired acceleration is to be achieved. In the case of an object being lifted vertically, the angle is 90°, of which the sine is 1, and therefore, the full weight of the object is the added force required beyond that which will provide the vertical acceleration. How much effort is needed to push a stalled car downhill? Obviously, it will be easier than pushing the same vehicle on a level road. In such situations, Eq. (4–35) will be changed to subtract $mg \sin \theta$.

Example 4–11. Consider a jogger weighing 136 lb who must accelerate 10 ft/s² to cross a busy street. This requires a force of the feet against the pavement of 42.5 lb. What force would be required if the jogger were to do this while running down a 15° hill?

Use Eq. (4–35):

$$
\begin{aligned}
F &= ma - mg \sin \theta \\
&= \frac{136 \text{ lb}}{32 \text{ ft/s}^2} \times 10 \text{ ft/s}^2 - 136 \text{ lb}(.2588) \\
&= 42.5 \text{ lb} - 35.2 \text{ lb} = 7.3 \text{ lb}
\end{aligned}
$$

What result would you get if the hill were 18°? 20°?

Example 4–12. Two shot-putters can each accelerate a shot 30 m/s². One applies force at an angle of 42°, while the other does so at 10°. What are the differences in force needed? (A 16-lb shot is 7.26 kg in metric units.)

Use Eq. (4–35):

$$F = ma + mg \sin \theta$$

At a 42° angle,

$$
\begin{aligned}
F &= 7.26 \text{ kg} \times 30 \text{ m/s}^2 + 7.26 \text{ kg} \times 9.8 \text{ m/s}^2 \times .6691 \\
&= 217.8 \text{ N} + 47.6 \text{ N} = 265.4 \text{ N}
\end{aligned}
$$

At a 10° angle,

$$F = 7.26 \text{ kg} \times 30 \text{ m/s}^2 + 7.26 \text{ kg} \times 9.8 \text{ m/s}^2 \times .1736$$
$$= 230.2 \text{ N}$$

UNIFORM CIRCULAR MOTION

In linear terms, when a body in motion covers equal distances in equal times, the body is in a state of uniform linear motion, and Newton's first law tells us that this is the natural condition when no unbalanced forces are present. Uniform circular motion exists when a body moves in a circular path over equal distances in equal intervals of time. In other words, the speed is constant but the *direction* of the body is continuously changing, and some restraining force must be operating to hold the body in this unnatural circular pathway. For example, a discus is held in a circular path by the grip of the fingers as the athlete spins around. When released, the discus leaves the hand and moves off in a linear manner tangentially to the circle it has been following.

There must be some physical restraint, a force that prevents a moving object from following a straight line and keeps it in a circular path. Circular motion is not the same as angular motion, because the object being restrained is not necessarily rotating around an axis within itself. By way of illustration, a person sitting on a carrousel horse is experiencing circular motion and is kept from going off in a straight line by maintaining a grip on the horse. The entire carrousel is undergoing angular motion as it turns on its own axis. One could argue that even the horse is also in circular motion only because it is being held in that path by bolts, and so it is apparent that this is a matter of defining terms in each situation. What constitutes the restraint and what is being restrained?

Centripetal Acceleration

Since linear velocity is a vector quantity and has both magnitude and direction, and since a change in velocity is by definition an acceleration, any change in the linear direction of a moving object means that there is an acceleration, which in turn must have been caused by an external unbalanced force. In cases of circular motion, this acceleration is directed toward the center of the circle along its radius and is called a *centripetal acceleration* or, in some texts, a *radial acceleration*.

This acceleration a_c is proportional to the square of the instantaneous tangential velocity and is inversely proportional to the radius of the circle. Thus,

$$a_c = \frac{v^2}{r} \tag{4-37}$$

where a_c = centripetal acceleration
v = tangential velocity
r = radius

Centripetal Force

In itself, centripetal acceleration, which is sometimes called radial acceleration, has little practical application in sports analysis except insofar as it is related to *centripetal force,* that is, the force that produces centripetal acceleration and the change of direction from linear to circular.

This force acts *on the object* in an inward direction along the radius toward the center, while the term more widely used in ordinary speech, *centrifugal force,* is nothing more than the equal and opposite reaction to the centripetal force and is the outward force exerted *on the restraint.* In a hammer throw, the cable is held tightly by the thrower's hands to keep the hammer in a circular path before release. There is a centripetal force acting on the hammer pulling it inward, and there is a centrifugal force acting outward on the athlete's hands. There is not an outward force acting on the hammer itself, because if there were, the hammer would take a path in line with the radius when released instead of at a right angle to the radius as it does (see Figure 4-7).

The calculation of centripetal and centrifugal force is done with Newton's second law, Eq. (4-1), with the value of centripetal acceleration from Eq. (4-37) substituted for acceleration. Thus, $F = ma$ becomes

$$F_c = \frac{mv^2}{r} \tag{4-38}$$

where m = mass moving in uniform circular motion
v = tangential velocity
r = radius
F_c = centripetal or centrifugal force

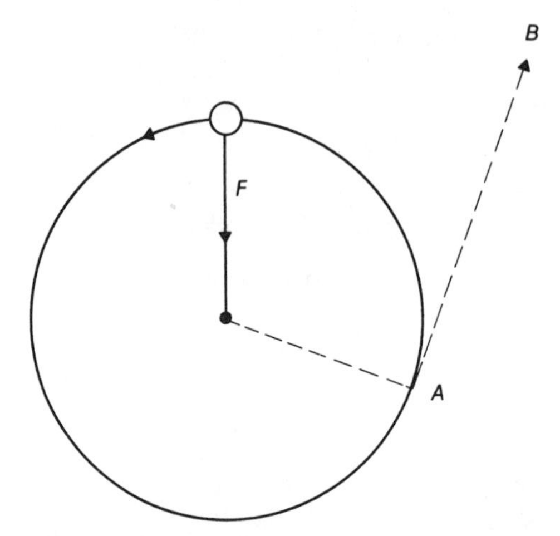

Figure 4-7 The hammer is held in a circular path by the tension F in the cable. If released at point A, the hammer will fly off in a straight line toward point B.

Example 4–13. A ball whose mass is 2 kg is twirled horizontally overhead at the end of a string 4 m long. If the ball's velocity is a uniform 2 m/s, what centripetal force does the string exert on the ball?

Use Eq. (4–38):

$$F_c = \frac{mv^2}{r} = \frac{2\text{ kg} \times (2\text{ m/s})^2}{4\text{ m}}$$

$$= 2\text{ kg}\cdot\text{m/s}^2 = 2\text{ N}$$

There are less obvious restraints that provide centripetal force, one of which is friction between the road and the tires of a vehicle negotiating a turn. If the road is not banked, the driver depends upon the friction factor to keep the car from sliding off the curve and must slow down if the friction seems insufficient. To help drivers get safely around tight highway turns, signs to urge speed reduction may be posted, or engineers can design the banking of curves to suit the speeds that most cars will be traveling around them. Correct banking takes into account the vertical component of weight down the slope that will provide the centripetal acceleration needed to offset the force tending to pull the car out of the curved path. Such banking cannot suit those driving more slowly or more rapidly than predicted by the design, nor will it work well if the road is wet or icy. Similar design problems exist for bobsled and roller coaster curves.

On a typical flat curve of an outdoor running track, as shown in Figure 4–8*a*, the centripetal force is provided by the runner's spikes digging into the track surface. The force acting inward on the feet would topple the runner outward if he did not lean inward enough to develop a gravitational moment that exactly balanced the outward reaction acting on his body. The angle of this lean from the vertical must be approximately the same as the angle to the horizontal that the track might have been banked had it been done so for the runner's specific velocity.

Many indoor running tracks are banked so that a runner can negotiate the turns while remaining essentially perpendicular to the track surface (see Figure 4–8*b*). The angle of lean from the vertical or the angle of banking to the horizontal that is necessary to maintain a desired speed around a curve with a given radius is found by the equation

$$\tan \theta = \frac{v^2}{gr} \qquad \textbf{(4–39)}$$

where θ = angle

v = expected velocity

g = gravity

r = radius of track curve

Note that the mass of the person or vehicle is not a consideration in Eq. (4–39). Also, when the curve is properly banked, the runner, in effect, goes round the turn at right angles to the track surface and therefore is not dependent on friction between his shoes and the track to provide centripetal force. Friction continues to be important for propulsive-force applications.

There is one problem with the otherwise neat Eq. (4–39), and it relates to the manner of determination of the radius of multilaned tracks. If for example the track has four lanes, each lane around a curve will have a different radius. The radius of the outside lane might be 8 ft more than that of the inside lane, and so the banking cannot be based upon the inside radius, because that would favor runners in that lane.

Another problem to be faced is that a curve banking designed for certain velocities may be less useful for races requiring other speeds. Small, indoor, noncompetitive tracks can be constructed to suit the jogger's speed as well as the sprinter's. A concave banking (Figure 4–9) can accomplish this, and the additional cost might be justified by the resulting benefits to the users. With Eq. (4–39), we can make some estimates of the banking depicted in Figure 4–9. If a jogger were to go around the curve at a leisurely 9 ft/s, this 30-ft-radius track at point *C* would be banked just under 5°. A miler running 18.7 ft/s would require a 19° bank at point *B* where the radius is 32 ft. The

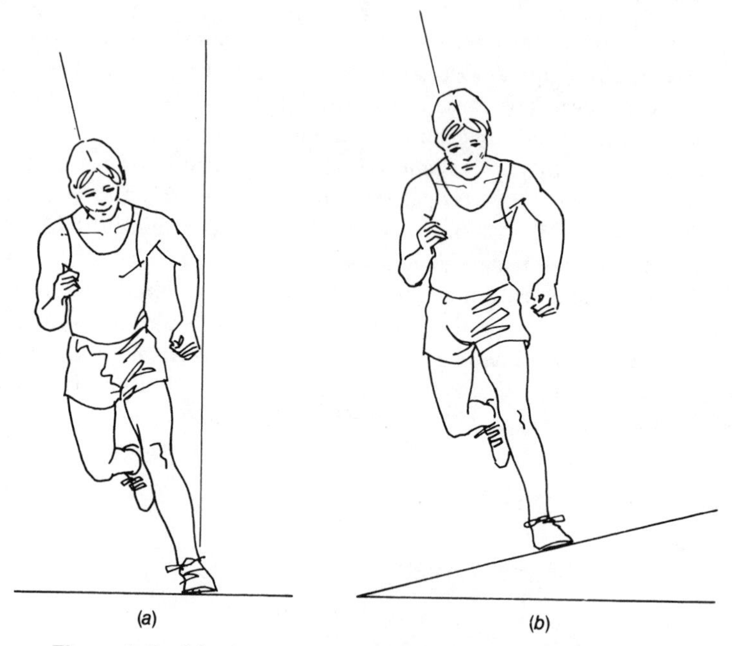

Figure 4-8 (*a*) A runner must lean from the vertical as he goes around a flat curve. (*b*) The track may be banked around a curve.

Figure 4-9 A cross-sectional view of a concave banked curve with radii of 30 ft, 32 ft, and 34 ft for points *C, B,* and *A* respectively. At *C,* a jogger making the turn at 9 ft/s will find the 5° bank suitable. Angles of 19° and 36° at points *B* and *A* accommodate speeds of 18.7 ft/s and 28 ft/s.

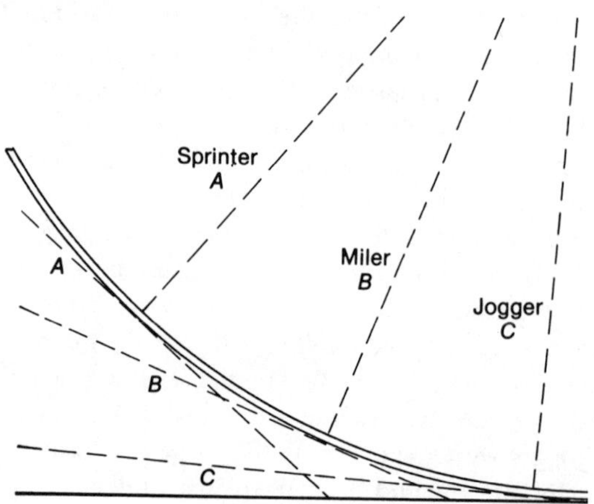

radius at point A is 34 ft, and a banking angle of 36° would be ideal for a sprinter moving at 28 ft/s around the turn.

Example 4–14. What should be the banking angle for a curve on an indoor track around which runners will have expected velocities of 29 ft/s? The radius of the curve is 34 ft.

Use Eq. (4–39):

$$\tan \theta = \frac{v^2}{gr}$$

$$= \frac{(29 \text{ ft/s})^2}{32 \text{ ft/s}^2 \times 34 \text{ ft}} = \frac{841}{1088}$$

$$= 0.7730$$

which is approximately the tangent of 38°, the angle of banking.

Example 4–15. A 175-lb runner goes around a *flat* curve at 25 ft/s If the radius of the curve is 50 ft,

1. How much centripetal force is applied?
2. How much must the runner lean inward?

1. Use Eq. (4–38):

$$F_c = \frac{mv^2}{r}$$

$$= \frac{175 \text{ lb} \times (25 \text{ ft/s})^2}{32 \text{ ft/s}^2 \times 50 \text{ ft}} = \frac{109{,}375}{1600} \text{ lb}$$

$$= 68.36 \text{ lb}$$

2. Use Eq. (4–39):

$$\tan \theta = \frac{v^2}{gr}$$

$$= \frac{(25 \text{ ft/s})^2}{32 \text{ ft/s}^2 \times 50 \text{ ft}} = 0.3906$$

which is approximately the tangent of 21°, the angle of lean from the vertical.

SUMMARY AND DISCUSSION

When a body's mass center moves in a straight-line path, its motion is classified as *rectilinear,* one of the two types of linear motion. The other, *curvilinear,* describes the movement of a body in a curved path as it is displaced from one point to another.

Newton's three laws of linear motion have been fundamental for nearly three hundred years, and their applicability to human movement has not been reduced by the discovery by Einstein and others that the laws do not fully work in the context of the universe and space travel. According to the laws, a body will remain in constant motion or at rest unless accelerated by an unbalanced force, and when such a force does act, the body will react in the opposite direction with an equal force. Problems in linear motion involve displacement, time, acceleration, and velocity. The velocity measured can be the instantaneous, average, final, or initial velocity, and it can be uniform or nonuniform. Acceleration may be uniform or nonuniform, but for practical reasons, this text assumes acceleration to be uniform.

Falling bodies are generally considered to be moving in a linear path, and the same equations apply to them as to bodies moving horizontally. Objects projected through the air are governed by the laws of linear motion and of falling bodies. Gravity is an ever-present force that brings unsupported bodies down to earth and is the unbalanced force that prevents a projectile from proceeding indefinitely in a straight line. As such it controls how high an object will rise, the vertical velocity of a projected object, and the time that the object will remain in the air.

Linear momentum, the quantity of motion, is found as the product of a moving object's mass and its linear velocity. The change in a body's momentum is a result of an *impulse,* which is defined as the product of the force applied and the duration of application.

Uniform circular motion of a body results from an outside restraint on the body's tendency to move in a straight line. The force that holds a body in a circular path is known as a *centripetal force.* Since it is directed inwardly, this force produces an inward acceleration known as a *radial* or *centripetal acceleration.* A running athlete who makes a change of direction must lean inward to keep from toppling outward, because a centripetal force acts on the feet to create a moment of force.

Solving the following problems might help the student to better understand linear motion and the related principles.

Problems for the Student

1. A 100-lb crate is being pushed by a force of 60 lb, which is resisted by 25 lb of frictional force. Find the box's acceleration.

2. A 192-lb barbell is lifted vertically and it accelerates at 1 ft/s^2. What force is being exerted?
3. A car moving at 60 mi/h must stop suddenly. The brakes are applied to give the car a uniform negative acceleration of 16 ft/s^2. How long will it take the car to stop and how far will it have traveled after the brakes are applied?
4. What is the mass of a body that is accelerated at 4 m/s^2 by an 80-N force?
5. What is the average force a 180-lb runner must exert horizontally in order to maintain a uniform acceleration of 16 ft/s^2?
6. A pole-vaulter clears the bar and falls 5 m to the pit. Find the impact velocity and the time it takes to fall.
7. A sprinter covers 18 ft during the first 1.5 s of uniform acceleration. What other data can be determined?
8. A field goal attempt has the result that the ball travels 99 ft in the air. The ball is in flight a total of 2 s from kick to landing. What else can be calculated about this kick?
9. How fast would a 6-oz ball have to travel to equal the momentum of a 16-lb shot moving 16 ft/s?
10. Find the impulse force for a ball kicked at a 40° angle from the ground, given the mass of the ball equal to 0.2 slugs, a contact time between foot and ball of 0.3 s, and a ball velocity of 100 ft/s.
11. An indoor track has a curve whose radius is 60 ft and whose banking angle is 18°. For what speed is such a banked track designed?
12. Discuss the potential dangers to a small boy's joints when an adult swings him by the hands in large circles parallel to the ground. What kind of force is acting?
13. From rest, a runner reaches a speed of 20 ft/s in 10 s. That speed is held constant for the next 20 s, after which the runner decelerates in 5 s to a stop.
    a. What are the two accelerations?
    b. What is the average velocity from start to stop?
    c. How far does the runner travel in that time?
14. What acceleration will be experienced by a 1-kg mass when it is acted upon by a 10-N force? What will be the velocity of the mass after 10 s? How far will it have moved in that time?

SUGGESTED READINGS

BUNN, JOHN, *Scientific Principles of Coaching,* 2nd ed. (Englewood Cliffs, N.J.: Prentice-Hall, Inc., 1972).

DYSON, GEOFFREY, *The Mechanics of Athletics* (London: University of London Press Ltd, 1967).

LEVIN, DAN, "Really Making Tracks," *Sports Illustrated,* Jan. 7, 1980, pp. 26–31.

MACMILLAN, M. B., "Determinants of the Flight of Kicked Footballs," *Research Quarterly,* vol. 46, no. 1, March 1975, pp. 48–57.

PLAGENHOEF, STANLEY, *Patterns of Human Movement* (Englewood Cliffs: N.J.: Prentice-Hall, Inc., 1971).

SCHVARTZ, ESAR, "Effect of Impulse on Momentum in Performing on the Trampoline," *Research Quarterly,* vol. 38, no. 2, May 1967, pp. 300–4.

TERAUDS, J., "Optimum Angle of Release for the Competitive Javelin As Determined by its Aerodynamic and Ballistic Characteristics," in R. C. Nelson and C. A. Morehouse, eds., *Biomechanics IV* (Baltimore: University Park Press, 1974), pp. 180–83.

TIPPENS, PAUL, *Applied Physics* (New York: McGraw-Hill Book Company, 1973).

CHAPTER 5

Angular motion

When movement takes place around an axis, it is termed *angular,* or *rotatory,* motion. The axis for such motion may be fixed and perhaps visible, as in the case of door hinges, or it may be free and perhaps invisible, as when a somersaulting diver rotates around an axis which passes through the body's center of gravity. Angular motion may be uniform or nonuniform. It is uniform when all parts of the rotating body turn through the same angle around the same axis.

Angular motion is initiated by the application of an unbalanced eccentric force, that is, by a force whose extended action line does not pass through the axis around which rotation is to occur.

For bodies that have fixed axes, the force vector tends to produce rotation whenever it passes through any point, even the center of gravity of the body, that is not serving as the axis. One of the most relevant examples of this condition might be that of the angular motion of any human extremity, because an eccentric force exists each time that a muscle contracts to move a limb around an adjacent joint. It is only through such angular movements of the limbs that a person can walk, run, kick, or throw.

For an airborne or an essentially unsupported body, rotation takes place in some plane around the body's center of gravity. Therefore, rotation will be produced by any unbalanced external force applied anywhere except through the mass center of such a free body. In the case of an airborne ball that is being struck, as in a volleyball serve, if the force vector is applied to the upper half of the ball and is directed above the ball's center of gravity, topspin will result, and the ball will also be given a linear acceleration in the direction of the applied force (see Figure 5-1).

Similarly, a bowling ball when first released may slide a short distance before the opposing frictional force acting on the ball's contacting perimeter will cause rolling to begin. The friction then can be classified as an eccentric resistance that acts on a body that otherwise would be moving in a linear manner according to Newton's first law. Another illustration of the effect of eccentric force or resistance is seen in the low tackling of a running football player, who falls around the axis created by the opponent's grasp.

MOMENT OF FORCE

As in the volleyball serve depicted in Figure 5-1, an eccentric force exists when the extended action line of an applied force passes some distance from an axis. The distance from the axis to the line of force is measured perpendicularly and is termed variously the *moment arm,* the *force arm,* the *effort arm,* the *lever arm,* the *resistance arm,* or the *radius.* Which expression is used depends upon the user's preference or upon the nature of the situation.

The product of an unbalanced force acting on a body and the moment arm is called a *movement of force,* and the unit is the pound-foot (lb•ft) or the newton-meter (N•m). A moment of force can be defined as the tendency to produce rotation, that is, motion around an axis. We say *tendency* because if the torque magnitude does not exceed the resistance afforded by the body, no rotation will occur. The term *torque* may generally be used synonymously with *moment of force,* but frequently the abbreviated *moment* will be used where a force has a turning effect, while *torque* is often favored to describe the twisting or torsion effect of a force. Moments are utilized whenever we open a door, turn a screwdriver, use a wrench, dig with a shovel, put on a hat—in other words, whenever we use all or part of any extremity.

In sports situations, moments may be internal, as when muscles contract to rotate limbs, or they may be external, as when gravity acts to create a

Figure 5-1 When the line of force passes above a ball's center of gravity, topspin results.

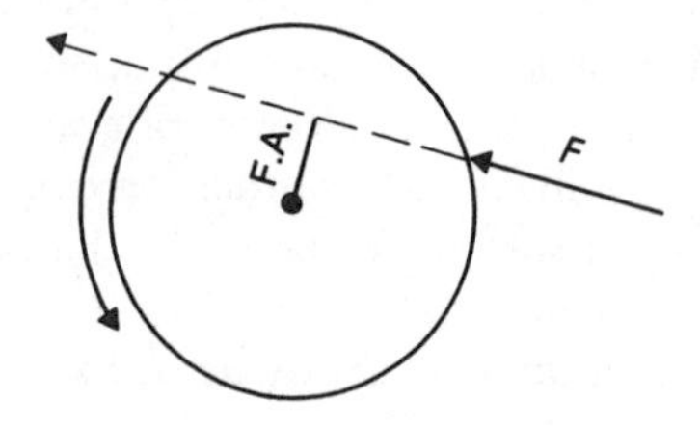

moment or when some implement strikes an object off center. Both of these types of moments can be illustrated by an athlete's kicking a ball off center to give it spin (Figure 5-2). The player's hip flexor muscles contract to move the femur from its slightly hyperextended preparatory position through an arc around the hip joint. This is closely followed by a contraction of the quadriceps femoris group to rotate the lower leg around the knee joint to make ball contact and follow through. The force of any group of muscles acting together can be expressed as some resultant force, which is then multiplied by a resultant moment arm, that is, the perpendicular distance from the center of the joint to the common line of pull of the muscle group, and a moment of force is thus obtained. These internal moments being nearly completed, the foot now contacts the ball, slightly above center in this case, and there exists a moment arm (F.A.), so that the ball will have topspin as well as linear motion.

The effectiveness of an unbalanced moment in producing rotation depends upon the magnitude of the force being applied relative to the resistance, the length of the moment arm, and the length of time of force application. Increased rotation can be obtained by either increasing the amount of force, lengthening the moment arm, or increasing the time of force application.

The line of force is diagramed as an extension of the force vector in either direction until it intersects the moment arm at a right angle. A moment of force or the torque can also be calculated by multiplying the applied force *F,* the distance *d* from the point of force application to the axis, and the sine of the angle formed by lines *F* and *d* (Figure 5-3). The full equation for moment of force or torque is then

$$\text{Torque} = Fd \sin \theta \tag{5-1}$$

Figure 5-2 Internal moments due to hip flexor and knee extensor muscle contractions cause the leg to rotate around the hip and knee joints. Topspin is imparted to the ball by the foot's contact above the ball's center of gravity.

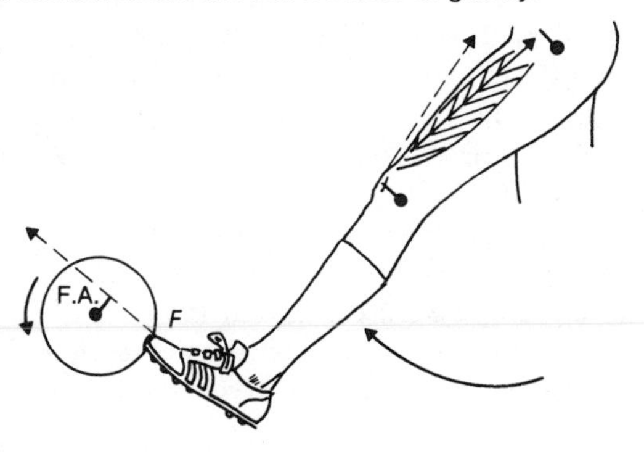

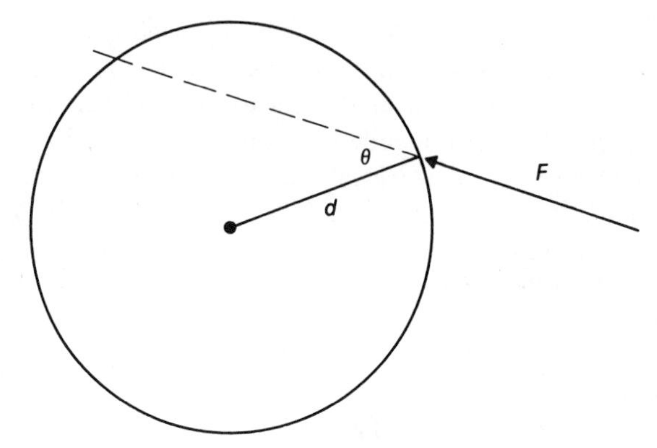

Figure 5-3 A moment of force can be calculated as the product of $Fd \sin \theta$.

A knowledge of moments is useful to athletic trainers and physical therapists, who use both isotonic and isometric exercises in working with the rehabilitation of muscle groups around joints. For instance, in the process of restoring the integrity of the knee joint after surgery or some injury, the knee extensor muscles can be progressively developed by the application of increasingly greater torques in opposition to the muscular torques. This can be done with particular types of equipment or by the application of manual resistance against the patient's efforts to extend the leg at the knee. To provide an initial small resistance when the quadriceps group is still weak, the trainer or therapist can place one hand slightly below the knee joint so that the resistance arm will be short and can further control the magnitude of the resisting moment by moderating the amount of force being applied by the hand (Figure 5-4).

As the athlete's muscles gain strength, the resisting torque can be gradually increased as the trainer's hand moves farther from the knee joint and as the applied force is increased. When the patient's leg is hanging vertically from the edge of the table, there is no moment, because the line of gravity passes through the knee and there is no moment arm. A simple, unresisted extending of the lower leg still finds opposition from the gravitational moment, which increases from a low value at the beginning to a maximum amount at full horizontal extension.

In recalling the muscle force discussion in Chapter 3, it is now possible to provide a more complete picture of what occurs when a muscle group contracts to rotate a limb. In Figure 3-8, only the angle of attachment of the tendon on the bone was of concern, and joint angle was not a factor. It should be evident that the degree of joint flexion does affect the line of pull of a mus-

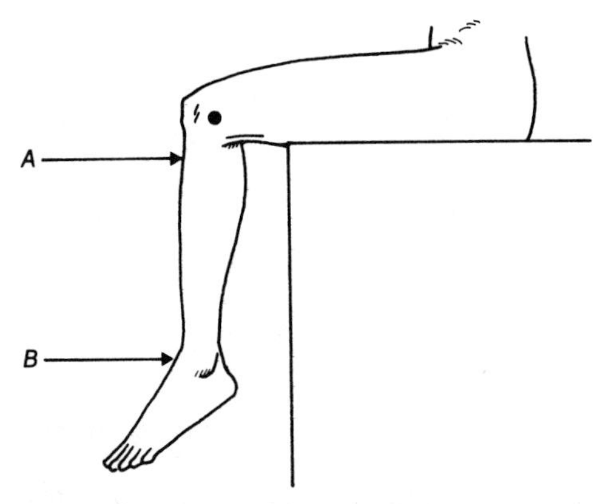

Figure 5-4 The moment of force resisting knee extension is greater at point *B* than at point *A*.

cle, which then in turn affects the moment arm length.

In the free body diagram shown in Figure 5–5, the moment arm is represented by the dotted line running from the center of the joint involved to the line of force of the muscle.

Example 5–1. Using the data in Figure 5–5, find the effective torque acting to rotate the limb.

If F_θ is 100 lb, the rotatory component of the force is $F_\theta \sin \theta$:

$$F_y = 100 \text{ lb} \times .766 = 76.6 \text{ lb}$$

If the moment arm is measured as 2 in., this is converted into feet and then multiplied by the rotational force component to obtain the torque. From Eq. (5–1),

$$\text{Torque} = \frac{2 \text{ in.}}{12 \text{ in./ft}} \times 76.6 \text{ lb} = 12.77 \text{ lb}\bullet\text{ft}$$

If SI units are to be used for this problem, 100 lb = 445 N, 2 in. = 5.08 cm, and 5.08 cm = 0.0508 m. Therefore,

$$F_y = 445 \text{ N} \times .766 = 340.9 \text{ N}$$
$$\text{Torque} = 340.9 \text{ N} \times 0.0508 \text{ m} = 17.32 \text{ N}\bullet\text{m}$$

This moment of force is valid only for this single angle with this amount of force acting. The student should appreciate that F_θ does not re-

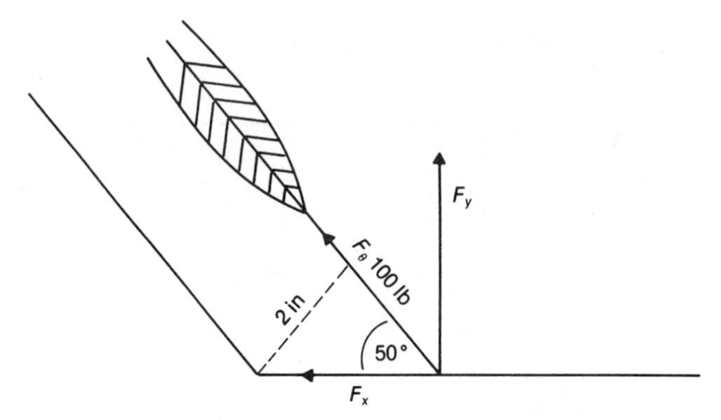

Figure 5-5 The perpendicular distance from the joint to the muscle's action line is shown as a dotted line. This moment arm, when multiplied by the applied force and the sine of the angle, gives the moment of force acting to rotate the segment.

main constant through the full range of motion of a limb and also that the force will vary as the velocity of muscle shortening varies.

ANGULAR MEASUREMENTS

In linear-motion conditions, it is easy enough to visualize a displacement of, say, 15 ft. Most of us have a reasonable idea of distances, although we may not all be able to envision metric distances readily. But in cases of angular motion, it would be meaningless to state that a body turns through a distance of 15 ft without somehow relating the movement to the circumference or radius of the rotating body. Different measurement units are required for the measurement of angular motion from those used for linear motion.

Angular Displacement

An angular displacement is the distance turned through or the amount of rotation expressed in terms of either radians, revolutions, or degrees, and the symbol for angular displacement is the Greek letter theta (θ). It can also be described as the product of the angular velocity and the time.

A revolution is, of course, one full circumference of a circle, which in turn is equal to $2\pi r$ or πD, r referring to the radius of the circle and D to its diameter. The Greek letter π (pi) denotes the ratio of the circumference of any circle to its diameter and has a value of 3.14. The revolution is the most commonly used unit of displacement in sports; a diver discussing a "forward one and a half'' is simply accustomed to omitting the unit, which is understood.

A degree is the angle subtended at the center of a circle by 1/360 times its circumference, and there are 360 degrees in any circle. This displacement unit is most often used in reference to range of joint motion or to vector angles but has a number of other uses, such as in compass readings.

A radian is the least familiar and least used unit in sports and physical education. It is a ratio of the length of an arc of a circle to its radius and as such is a dimensionless number. The radian (rad) is defined as an arc of circumference whose length is equal to the radius of the circle. Thus, if a circle has a radius of 10 in., a radian is a 10-in. segment of the circumference. Since there are 2 π rad in a circle, the circumference in this case is $2 \times 3.14 \times 10 = 62.8$ in. When the number of degrees in a circle is divided by the number of radians in a circle, the result is that 1 rad = 57.3°.

A little practice in the use of Table 5-1 should make the relationships between the three displacement units clear.

Example 5-2. How many degrees of angular displacement are represented by 3.2 radians?

$$3.2 \text{ rad} \times 57.3°/\text{rad} = 183.4°$$

Example 5-3. How many revolutions are equivalent to 4.6 radians?

$$4.6 \text{ rad} \times 0.159 \text{ rev/rad} = 0.73 \text{ rev}$$

Example 5-4. How many radians are there in 3 revolutions?

$$3 \text{ rev} \times 6.28 \text{ rad/rev} = 18.84 \text{ rad}$$

Example 5-5. How many degrees are in 6.5 revolutions?

$$6.5 \text{ rev} \times 360°/\text{rev} = 2340°$$

TABLE 5-1
Conversions between Units of Angular Displacement

Given	*To Obtain*	*Multiply by*	*or Divide by*
Radians	degrees	57.3	.01745
Degrees	radians	.01745	57.3
Radians	revolutions	.159	6.28
Revolutions	radians	6.28	.159
Degrees	revolutions	.002777	360
Revolutions	degrees	360	.002777

Angular Velocity

The angular velocity for a body rotating at a uniform rate is determined by dividing the angular displacement by the time and is defined as the rate of change of angle. The symbol for angular velocity is the Greek letter omega (ω).

$$\omega = \frac{\theta}{t} \tag{5-2}$$

The resulting units will be either radians per second, revolutions per second, or degrees per second, depending on the units given for the angle θ.

Angular Acceleration

An angular acceleration results from an unbalanced torque acting upon a body. It may be uniform or nonuniform, and it is represented by the Greek letter alpha (α). Angular acceleration is defined as the rate of change of angular velocity and is determined by the equation

$$\alpha = \frac{\omega_f - \omega_o}{t} \tag{5-3}$$

The resulting units will be either in radians per second per second, revolutions per second per second, or degrees per second per second.

Linear Velocity

The linear velocity of any point along the radius of a rotating rigid body represents the linear speed with which that point will travel in a linear tangential direction if it is released from its angular path at a given instant. A rotating body has an angular velocity that is the same for all parts of that body, but any point on any radius also has a linear velocity that is a function of its distance from the axis. The linear velocity of a point is the product of its angular velocity and its distance along the radius to the axis:

$$v_L = \omega r \tag{5-4}$$

where r is the length of the radius from the point to the axis, v_L is expressed in feet per second or meters per second, and ω is expressed in radians per second.

$$v_L = 2\pi Nr \qquad (5\text{–}5)$$

where r is the radius, v_L is in feet per second or meters per second, and N is expressed in revolutions per second.

Linear velocity is commonly expressed in feet or meters per second, and the concept might best be illustrated by a line of marchers executing a turn (Figure 5–6). Because the entire disciplined line moves as a unit through a given angle in a given time, it behaves as would a solid rod or radius, and each marcher in the line can be said to have the same angular velocity. However, it is obvious to anyone who has observed or been part of such a maneuver that the person most distant from the pivot must move much more rapidly than those who are closer to the pivot. That the more distant person's linear velocity is greater can be confirmed by applying Eq. (5–4) to each person in the line.

Example 5–6. One line of a marching band makes a turn, as in Figure 5–6, and goes through a 60° angle in 10 s. If member A is 30 ft from the pivot and member B is 70 ft from the pivot, what are the linear velocities of these two marchers?

Use Eq. (5–2):

$$\omega = \frac{\theta}{t} = \frac{1.047 \text{ rad}}{10 \text{ s}} = 0.1047 \text{ rad/s}$$

Figure 5-6 When a marching line makes a turn from *OV* to *OW*, all members take the same time to reach *OW*, but those farthest from *O* must move the fastest.

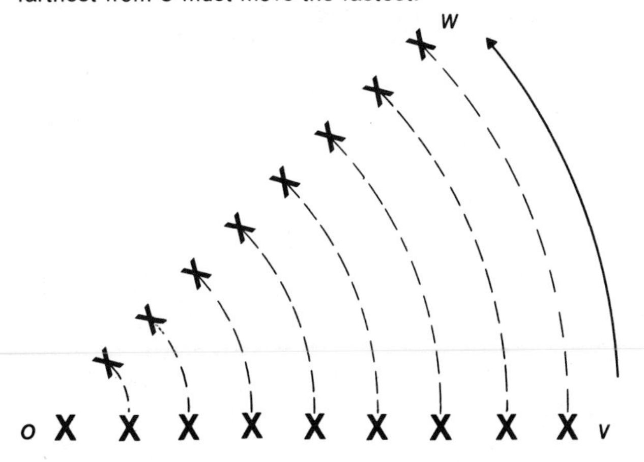

(1.047 rad was obtained from Table 1 by multiplying 60° by 0.01745.)

Use Eq. (5–4):

$$v_L = \omega r$$

$$v_L = 0.1047 \text{ rad/s} \times 30 \text{ ft/rad} = 3.141 \text{ ft/s} \qquad \text{for } A$$

$$v_L = 0.1047 \text{ rad/s} \times 70 \text{ ft/rad} = 7.329 \text{ ft/s} \qquad \text{for } B$$

Alternatively, angular velocity could have been expressed in revolutions per second, in which case Eq. (5–5) would have been used to obtain linear velocity.

To summarize, the linear velocities of various points along the radius of a rigid rotating body are proportional to their respective distances from the axis. This concept is especially important in those sports where an angular velocity is developed in order to release or strike some object at a maximum velocity (see Figure 5–7). As one example, a discus thrower needs to develop a high angular velocity in movement across the ring and has to try to release the discus from the longest radius that is manageable. The combination of high angular velocity and long radius will provide the best linear velocity for the discus at release. In baseball batting, the outer third of the bat, where contact is normally made, will have a higher linear velocity than either the middle or the proximal third when the bat's angular velocity is held constant. When the outer third strikes a baseball, the best potential linear speed is achieved, other things being equal. A number of other factors are involved in batting, and the only intention here is to relate linear velocity, angular velocity, and radius.

Example 5–7. A point on a radius lies 1.5 m from the axis and rotates at the rate of 50 revolutions every 2.5 seconds.

1. What is the angular velocity of the radius and therefore of the point?
2. What is the linear velocity of the point?

1. From Eq. (5–2):

$$\omega = \frac{\theta}{t}$$

In revolutions per second,

$$\omega = \frac{50 \text{ rev}}{2.5 \text{ s}} = 20 \text{ rev/s}$$

Figure 5-7 In the tennis serve, a fully extended elbow provides a high linear velocity for the racket head. *Photograph courtesy of the Ohio State University Athletic Publicity Department.*

In degrees per second,

$$\omega = \frac{50 \text{ rev} \times 360°/\text{rev}}{2.5 \text{ s}} = 7200°/\text{s}$$

In radians per second,

$$\omega = \frac{(2 \times 3.14) \text{ rad/rev} \times 50 \text{ rev}}{2.5 \text{ s}} = 125.6 \text{ rad/s}$$

2. From Eqs. (5–4) and (5–5),

$$v_L = \omega r = 125.6 \text{ rad/s} \times 1.5 \text{ m/rad} = 188.4 \text{ m/s}$$

$$v_L = 2\pi N r = \frac{(2 \times 3.14) \text{ rad/rev} \times 50 \text{ rad} \times 1.5 \text{ m/rad}}{2.5 \text{ s}} = 188.4 \text{ m/s}$$

MOMENT OF INERTIA

In Chapter 4, it was stated that the inertia of a body is proportional to the body's mass and that inertia is a measure of a body's resistance to any change in its state of linear motion. In cases of angular motion, the equivalent term for inertia is *moment of inertia* (I). The moment of inertia of a rigid body which is rotating, or is capable of being rotated, is a measure of its opposition to any change in its state of angular motion.

Unlike linear inertia, moment of inertia is concerned less with the total mass of a body than with how that mass is distributed around the axis of rotation. Equations to determine I vary according to the shape of the object, but generally the moment of inertia for the human form is obtained by summating the products of all the mass particles of the body and the squares of their respective perpendicular distances to the axis. It may be expressed as:

$$I = \Sigma \, mr^2 \qquad \textbf{(5–6)}$$

The unit is slug•ft^2 or kg•m^2.

It is sometimes more convenient to use the expression *radius of gyration* (k), which is defined as a radius drawn from the axis to some point that is at the average distance of all the mass particles from the axis. Eq. (5–6) would then be replaced by

$$I = mk^2 \qquad \textbf{(5–7)}$$

where m represents the total mass. If the moment of inertia is known, Eq. (5–7) can be rearranged to solve for k:

$$k^2 = \frac{I}{m}$$

In many sports, particularly gymnastics and diving, the radius of gyration and even the axis change frequently, and the moment of inertia changes accordingly. A coach or physical education teacher would rarely attempt an accurate mathematical determination of moment of inertia for any athlete or student, but the concept and its effects on performance must be understood by both the teacher and the participant.

A good example of this change in radius length and moment of inertia is the difference in arm flexing in walking as compared with running. The arm is pretty much fully extended and relaxed when a person walks, and therefore the moment of inertia of the arm is maximal as it rotates around the shoulder joint. Since the movement is relatively slow, the high moment of inertia is of no concern. During a period of running, the arms must be swung more rapidly to keep up with the increased stride frequency, and so it becomes necessary to flex the arms at the elbow to reduce their moments of inertia and hence their resistance to rotation. In flexion, the mass particles of each arm are brought closer to the shoulder joint axis, and this makes it easier for the shoulder muscles to rapidly move the arms.

When dealing with limb movement, one must take care not to confuse the moment of inertia of the limb with the additional resistance to motion afforded by gravity in certain movements. Consider an exaggerated arm swing during walking. The forward-upward phase of the swing would seem more difficult than the downward-rearward phase, since both the force of gravity and the arm's resisting moment of inertia are acting. Because of the help from gravity, the return swing seems easier, but the moment of inertia itself remains the same during both the forward and rearward swings. Similar relationships between gravitational force and moment of inertia could be cited for a variety of arm or leg actions during exercise or skill performance. Gravity is not a factor in movements made in the transverse plane, at least not in terms of resistance to change in the state of angular motion in that plane.

At best, calculating the moment of inertia for an irregularly shaped object such as the human body, even when it is motionless, is a complicated task. Because of the difficulty in measuring moments of inertia of the human body or of its separate segments during a sports performance, it will suffice for purposes of this text to discuss only the *effects* of changes in the moment of inertia. Students interested in a fuller understanding of moment of inertia and the methods for its determination may read Plagenhoef[1] or Miller and Nelson,[2] who provide good treatment of the topic.

NEWTON'S LAWS OF ANGULAR MOTION

The laws of motion with which most students are familiar are those which relate to linear motion, and these were discussed in Chapter 4. The application of these laws to angular motion requires their rephrasing and the use of

[1] Stanley Plagenhoef, *Patterns of Human Motion* (Englewood Cliffs, N.J.: Prentice-Hall, Inc., 1971).

[2] Doris I. Miller and Richard C. Nelson, *Biomechanics of Sport* (Philadelphia: Lea & Febiger, 1973), pp. 89–90.

angular terminology. Note that the term *torque* may be substituted wherever *moment of force* is used below.

The First Law of Rotational Motion

A rigid rotating body will continue to revolve about an axis, fixed in direction, with constant angular velocity unless acted upon by some external unbalanced moment of force.

As in linear motion, the ever-present external forces that act to slow or stop angular motion are friction and air resistance and, in some cases, gravity. A rotating or resting body's moment of inertia is a measure of its resistance to changes in its state of angular motion around a particular axis.

The Second Law of Rotational Motion

If an unbalanced moment of force acts upon a rigid body, an angular acceleration is produced which is proportional to the moment of force and inversely proportional to the moment of inertia of the body.

This law also applies to bodies that are being decelerated, and a rotating object that has either a high angular velocity or a high moment of inertia or both will require a large moment of force to slow or stop it.

The Third Law of Rotational Motion

For every moment of force acting on one body, there is an equal and opposite moment of force about the same axis which acts upon some other body.

In other words, for every angular action there is an equal and opposite reaction, or for every torque there is an equal and opposite counter-torque. This law has particular significance for unsupported bodies undergoing angular motion.

LAW OF CONSERVATION OF ANGULAR MOMENTUM

The law of conservation of angular momentum relates directly to the first and third rotational laws and states that

In the absence of an external unbalanced moment of force, a rotating body has an angular momentum which is constant in both magnitude and direction.

A body at rest has no angular momentum and will continue to have none unless some torque produces an acceleration. Also note in the definition that once angular momentum is established, the axis direction will not change.

Angular momentum may be defined as the product of a body's moment of inertia and its angular velocity, or

$$A = I\omega \tag{5-8}$$

The unit is a rather clumsy slug•ft² rad/s or kg•m² rad/s.

Since the angular momentum of a freely rotating body has to be conserved, in the absence of external moments, any change in the moment of inertia of the body will bring about an opposite proportional change in its angular velocity. If an athlete who is freely revolving decreases his or her moment of inertia by bringing body segments closer to the axis, the body's angular velocity will increase accordingly to keep the angular momentum constant.

Rotating body segments are subject to this law, but a segment is seldom free to rotate, since it is under the control of muscles that have attachments outside the segment. Thus an external force due to muscular action and sometimes due to gravity as well is always acting during the major portion of a segment's angular motion. So the law of conservation of angular momentum is generally not applicable to analysis of extremity movements. As one illustration, the recovery phase of the leg in sprinting involves considerable flexion at both the knee and the hip joints, and as a result the leg has a higher angular velocity than it would if it were less flexed. This higher velocity is due to the reduced moment of inertia of the leg, which permits the hip flexor muscles to move the leg more rapidly because of the decreased resistance to change in angular motion. The muscles provide an external moment of force, and the leg is not freely swinging, which rules out the applicability of momentum conservation.

A gymnast performing on a horizontal bar, as in Figure 5–8, is continuously being acted upon by gravity, which constitutes an external force, but some aspects of his swinging movements are free enough to allow us to make use of the conservation law. The gymnast develops an angular momentum and then manipulates his radius and moment of inertia, shortening it to increase his angular velocity when he must rise higher against gravity and lengthening it on the way down to obtain the best turning moment from gravity.

More correct applications of the law of conservation of angular

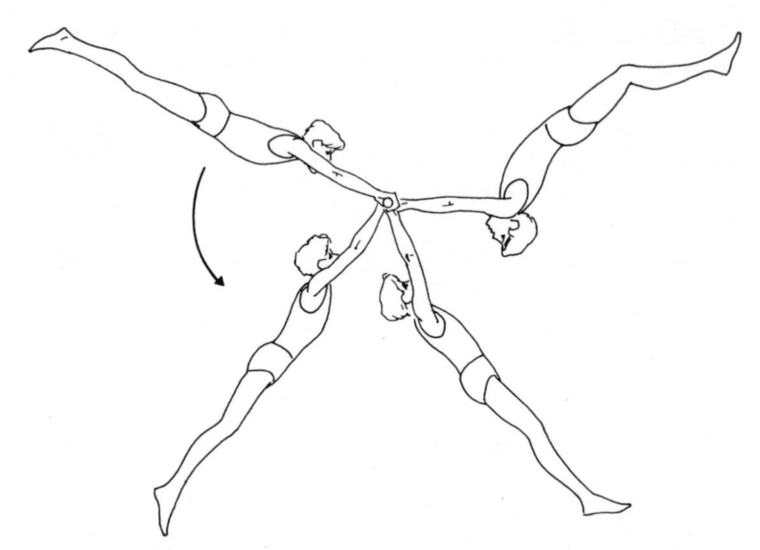

Figure 5-8 On the horizontal bar, a gymnast increases his downswing speed by fully extending his body. He increases his angular velocity by shortening his radius on the rise.

momentum can be made to those sports skills in which an athlete is in the air, as in diving or trampolining, or is on a fairly friction-free surface, as in ice skating. In such sports, changes in moment of inertia are accompanied by changes in angular velocity, so that total angular momentum remains the same.

As a man dives from a springboard, he has some angular momentum while in an extended layout position. If he then assumes a tuck position, thereby reducing his moment of inertia perhaps threefold, his angular velocity will increase on the order of three times. The only part of Eq. (5-8) under his control is I, since, being unsupported, he cannot by any internal force actions directly affect his angular velocity. The axis for rotation passes through his center of gravity, and in most diving skills, it is either a lateral or a longitudinal axis.

A woman figure-skating makes a semicircular approach to the point of spin, brings the free leg around in a wide arc to create a high angular momentum, and begins to spin. Her angular velocity can be controlled by the manner of distribution of her mass around her longitudinal axis. With abducted arms and free leg, rotation will be slow. If the arms and free leg are adducted in any manner, the moment of inertia will be reduced and the angular velocity will be increased. This principle also applies to jump-turns. If more than one revolution is desired, the extremities must be brought inward to speed the spin.

In the earlier discussion of linear velocity of a point on a rotating radius, it was noted in Eq. (5–4) that v_L was the product of angular velocity and the radius. There is a relationship here to Eq. (5–8) in that a moment of inertia includes a radius measure and, while I and r are certainly not the same, both have a similar inverse association with angular velocity. If the linear velocity of a point on a freely rotating body is, say, 20 ft/s for a given angular velocity and length of radius, it will continue to be 20 ft/s if the radius is reduced, thereby increasing the angular velocity. It can be stated then that the linear velocity of a point is conserved or is constant for any freely revolving body regardless of changes in radius length.

To illustrate the implications of this principle, consider a discus thrower, whose angular motion is affected by his moment of inertia. Prior to releasing the discus, he develops his highest possible angular momentum. Since for the most part he is essentially unsupported through his turns, the law of conservation of angular momentum applies. What might he gain by drawing in his arms during the turn before release? He should spin faster as a result, but will the throw be improved? Remember that the linear velocity of the discus is a function of angular velocity and radius. It will do no good to increase the angular velocity while reducing the radius; the linear velocity will remain the same. The detrimental effect of such a maneuver might be loss of balance because of the faster rotation.

There are several ways in which the law of conservation of angular momentum may be demonstrated in the classroom, of which the simplest is the use of a free-turning stool.

1. Sit on the stool with the feet clear of the floor and with the arms held horizontally abducted. Vigorously swing both arms toward each other (horizontal flexion). Note that there is no "reaction" in the rest of your body, nor is there any movement of the stool itself, because the oppositely moving arms serve as the reactions to one another and the total angular momentum, which was zero before, remains zero (see Figure 5–9).

2. From the same position as in demonstration 1, now vigorously swing (horizontally flex) only one arm, and note that the rest of your body and the stool will rotate in a direction opposite that of the swinging arm (see Figure 5–10). Here again, the original angular momentum was zero and had to be conserved; therefore the magnitude of the body-stool reaction required is that product of moment of inertia and angular velocity which will exactly offset the angular momentum of the swinging arm.

This demonstration explains why it is so difficult for an airborne athlete to initiate twists or other rotations after leaving a supporting surface. Although, given time and skill, such rotations can be initiated in the air, it is much more effective to begin rotation while still in contact with the supporting surface. Note, however, that an athlete who already has an angular

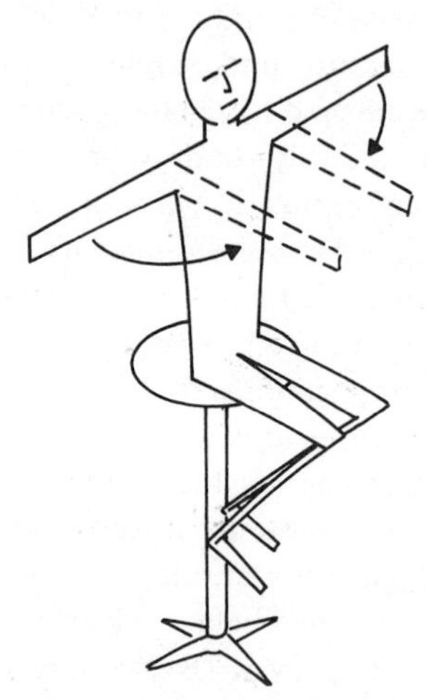

Figure 5-9 Angular momentum will remain zero if two equal segments move toward one another, each serving as a reaction to the other.

Figure 5-10 As an arm horizontally flexes, the rest of the body will turn in the opposite direction, so that total angular momentum remains zero.

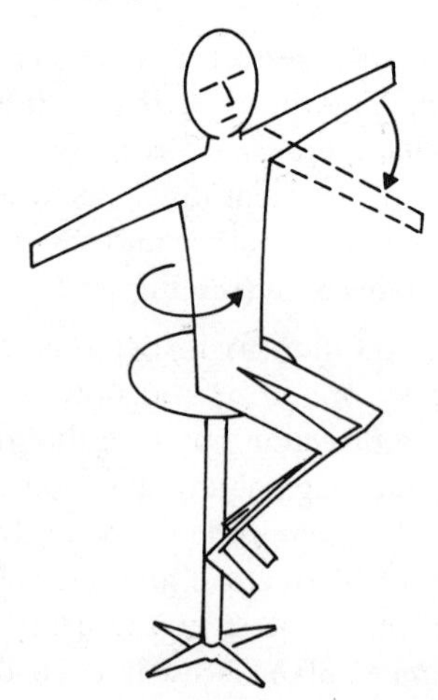

momentum can alter the axis of rotation as, for example, from a somersault to a twist. This has been termed *trading* angular momentum about one axis for angular momentum around another.

3. While sitting on the stool as before, increase your moment of inertia by extending your legs at right angles to your trunk. Now repeat the arm swing as in demonstration 2 and note that the reaction is somewhat less than it was there. The reason for this is that the higher moment of inertia due to the leg position means that a relatively low angular velocity will be enough to provide the reaction needed to maintain the zero angular momentum. The amount of friction in the stool will determine how well these demonstrations uphold the law of momentum conservation.

4. With the feet on the floor and the arms held in an abducted position, push off with the feet so as to set your body and the stool into rotation and then lift both feet clear of the floor. An angular momentum is now established and it will be conserved regardless of changes you make in your moment of inertia. If the arms are quickly folded close to the body, there will be a sharp increase in the angular velocity of your body. If the arms are now reextended, the angular velocity will be reduced. This demonstration is even more dramatic if you hold small weights in your hands.

TRANSFER OF ANGULAR MOMENTUM

The same type of stool may be used to show a variation of the law of conservation of angular momentum, that is, transfer of angular momentum from a part to a whole. Keep one foot, say the left, on the floor to absorb reaction and then vigorously swing the free right leg in a counterclockwise direction. Just at the point where the right leg has developed its maximum angular momentum, raise the left foot from the floor. The entire body will then rotate counterclockwise because of the transfer of momentum from the right leg to the whole body. In other words, the angular momentum of the right leg will be conserved. The timing of the left-foot lift is critical, since lifting it too soon will result in a clockwise reaction, and raising it too late will reduce or eliminate transfer effects because the ground will have absorbed the momentum.

The moving body part must *lock* onto the rest of the body by a sudden fixing of the joint around which it is moving. This is the principle used by a person who is lying supine and sits up by first flexing at the hips to get the legs vertical and then sharply extending at the hips. When the legs are about halfway back down, the antagonistic hip flexor muscles contract to brake the swing, thus causing the previously resting upper body to assume the angular momentum and to follow the legs to a sitting position (see Figure 5–11).

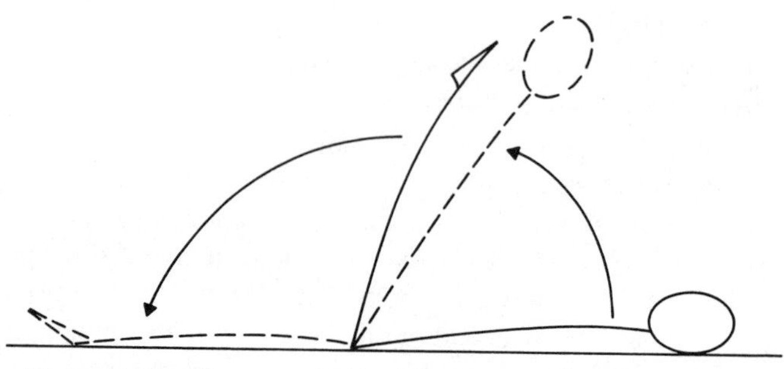

Figure 5-11 From a supine position, a person can rise to a sitting position (dotted lines) by first elevating the legs (solid lines) and then swinging the legs back toward the floor but locking the hip joint after beginning the swing.

It is also possible to transfer angular momentum from the whole to a part, as when a rapidly spinning discus thrower slows body rotation just at the instant before release, causing the throwing arm to accelerate. Whenever the more massive segment decelerates, a smaller distal segment that is free to move will experience a positive angular acceleration (see Figure 5-12). This also occurs in kicking movements, where the knee of the kicking leg is kept well flexed until the point where the thigh begins to slow.

Figure 5-12 On uneven parallel bars, the momentum developed in the downswing is suddenly checked by the lower bar, resulting in an acceleration of the legs.

A number of analogies between linear and angular terms and principles have been presented. Table 5-2 brings most of these together for reference.

LEVERS

Of the several simple machines such as the wheel and axle, the pulley, the inclined plane, and the jackscrew, the lever is the one with the widest relevance. A lever may be defined as a rigid bar that is used to overcome a resistance when a force is applied to one side of a fulcrum.

Levers can be found either internally in the form of extremity bones or externally in the form of sports implements such as rackets, bats, and poles. It should be made clear that the bar referred to in the definition can be of any shape.

There are three types or classes of levers, and the student should commit these schematic diagrams to memory (see Figure 5-13).

In levers of the first class, the fulcrum may be moved about along the lever, thereby changing the relative lengths of the force arm and the resistance arm. If the fulcrum is placed close to the resistance, the force arm is lengthened and less force need be applied to move the resistance, but the force must be applied through a long distance in order to lift the resistance a short distance. Conversely, a shortened force arm requires greater force application, but there is a gain in speed and range of motion at the resistance end.

Movement of the fulcrum in second-class levers will increase or decrease both the force arm and the resistance arm. The force arm is always

TABLE 5-2
Some Linear Quantities and Their Angular Analogues

Linear		*Angular*	
Term	*Symbol*	*Term*	*Symbol*
Displacement	d	Angular displacement	θ
Velocity	v	Angular velocity	ω
Acceleration	a	Angular acceleration	α
Force	F	Moment of force, torque	L
Mass	m	Moment of inertia	I
Momentum	P	Angular momentum	A
Second law	$F = ma$	Second law	$L = I\alpha$
Impulse	Ft	Angular impulse	Lt
Work	Fd	Angular work	$L\theta$

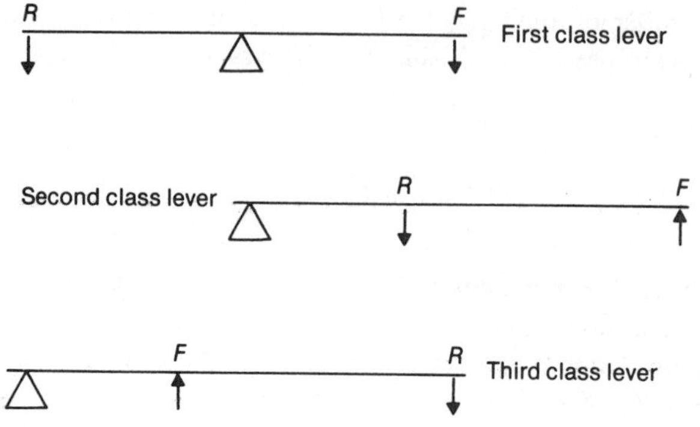

Figure 5-13 The three classes of levers shown schematically

the longer of the two, and therefore the force needed to lift a resisting weight will always be less than the weight.

The force arm of a third-class lever is always shorter than the resistance arm, and so a large amount of force must be applied, but the resistance is moved through a much longer range of motion than that of the force application. In the human body, the most common class of lever is the third, and this is particularly important in the motions of the limbs, because the results desired are very often those of speed or range of motion, albeit at the expense of force. Figure 5-14 depicts the abduction of an arm by the middle deltoid and supraspinatus muscles, whose distal attachments on the humerus represent the points of force application. The center of the shoulder joint is the axis, and the resistance is the combined weights of the arm and whatever the hand might be grasping. The force arm, if it could be measured in a living human, is the perpendicular distance from the joint center to the line of pull of the muscles. The resistance arm is the perpendicular distance from the joint center to a vertical line passing through the center of gravity of the segment plus whatever is being held.

If the arm is stationary, there is a static contraction of the involved muscles; that is, the muscles apply only enough force to exactly resist the downward moment of force acting on the arm. If further abduction occurs, it is because there is a net muscular torque, and a small amount of motion in these muscles causes a much longer arc of movement at the end of the arm.

Since both the force and the resistance act at some perpendicular distance to the axis, they constitute moments of force. When the lever definition states that a force is used to overcome a resistance, it really means that the magnitude of the moment of force (force times force arm) is greater than the resisting moment of force (resistance times resistance arm) and that rotation will occur about the axis in the direction of the larger moment.

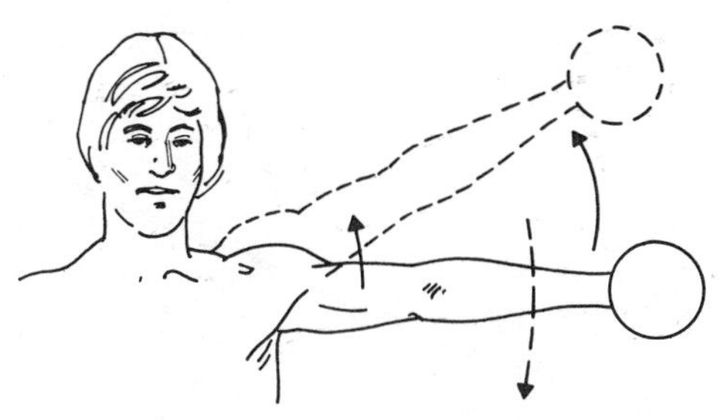

Figure 5-14 An unbalanced torque due to contraction of the abductor muscles causes further abduction.

The relationship between opposing moments is represented by the equation:

$$F \times FA = R \times RA \quad (5\text{-}9)$$

where F and FA are the force and force arm and R and RA are the resistance and resistance arm, respectively. Both sides of the equation are equal, and so the sum of the two moments acting in opposite angular directions is zero and a condition known as *equilibrium* exists and represents a study in statics. Note that the term *equilibrium* is used to identify any condition in which the sum of the forces or torques acting equals zero, and so it can be applied to both static and dynamic conditions of linear or angular motion.

It is essential to remember that where gravity is acting to cause a moment, the line of gravitational force from which the force arm is to be measured is always a vertical line. If the third-class lever diagram from Figure 5-13 is inclined somewhat, as in Figure 5-15, the resisting gravitational force line is drawn vertically and its moment arm is drawn at right angles to connect it with the axis. The line of force is directed to resemble the muscular pull shown in Figure 5-14; its moment arm is labeled FA. It is a common error for students to think of a resistance arm as simply the length of the lever itself between the center of gravity of the segment and the axis. This is true only when the lever is horizontal, at which time it is itself perpendicular to the gravitational line of force.

The definition given for center of gravity in Chapter 1 can now be reworded to state that the mass center of a body is a point around which all the torques summate to zero. The body is in equilibrium when suspended in any position from that point. A simple example is the placement of a finger under the midpoint of a yardstick. The positively and negatively acting moments balance each other, and the location of the center of gravity of the stick is confirmed. Such balance is passive, but in an erect human posture, as in standing at attention, the equilibrium condition is not passive, because

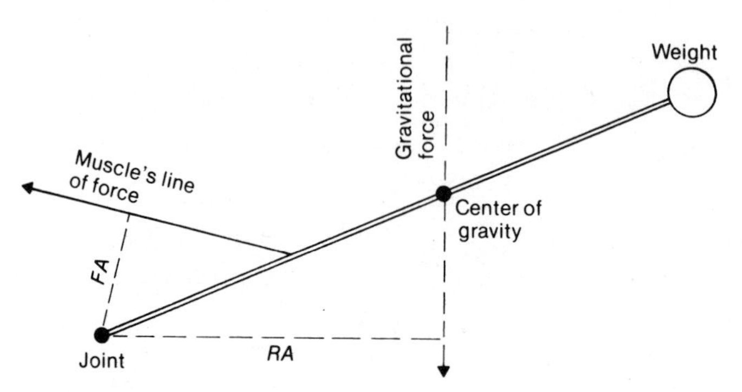

Figure 5-15 In a body segment representing a third-class lever, the force arm and the resistance arm (dotted lines) are shown drawn perpendicular to the two forces that are involved.

muscular forces are continually acting on segments to hold a steady posture. All segments are not vertically aligned, and so there are constant unbalanced torques appearing and then being instantly counteracted by actions of the postural muscles to make it appear to the casual observer that there is no movement of the body.

While there is no interest in this text in classifying the numerous levers within the body, most of which are of the third class, it might be profitable practice for the student to identify levers used in some common movements and exercises, as shown in Figure 5-16 and as might be found in the student's own sports specialty. It is not unusual to be able to identify more than one moment acting upon even a single extremity, and so care must be taken to examine every joint.

STABILITY

While seemingly out of place in a chapter on angular motion, the topic of stability probably belongs here because of its association with equilibrium and because instability results when a turning moment causes the body to fall or rotate about one edge of its base.

When we say something is stable, we generally mean that it is not easily upset, that is to say, that it takes some effort to topple it. By contrast, of course, an unstable object is one that is easily upset.

A *line of gravity* is defined as an imaginary vertical line that passes through a body's mass center. An object will topple if its line of gravity falls outside its base of support. In the human, the center of gravity, and hence the line of gravity, may move about relative to the support base, but it must fall

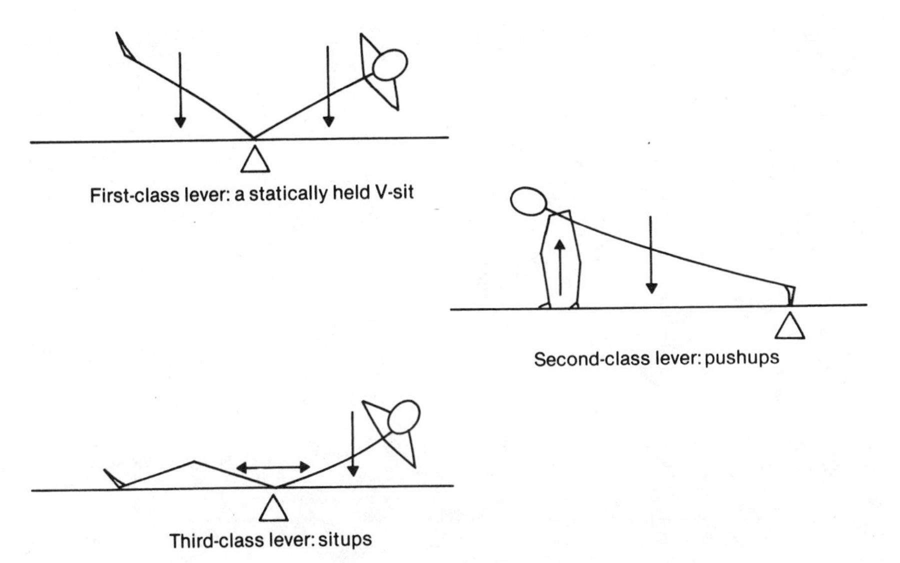

Figure 5–16 Examples of levers used during exercises.

within the boundary limits of the base if a stable condition is to exist. There are two types of stability: static and dynamic.

Static Stability

As its name indicates, static stability involves held positions. An athlete may be momentarily motionless either in preparation to perform a skill or in performing part of the skill itself. A gymnast holding a handstand position (as in Figure 5–17) offers an example of the former condition, while the latter condition is represented by the set position in track and swimming starts. Static stability is very important in such skills as archery and shooting and in a number of gymnastic stunts.

Stable objects generally have wide bases and low centers of gravity, and unstable objects have just the opposite characteristics. A third category is that of objects in neutral equilibrium, in which the height of the center of gravity is unaffected by a push, as in the case of a ball lying on the floor.

Achieving optimum static stability depends on one or more of the following principles:

1. The larger the area of the base of support, the greater the stability. If a person widens his or her stance or gets down on hands and knees, the base is widened and stability is improved.

Figure 5-17 An example of static stability on the balance beam. *Photograph courtesy of the Ohio State University Athletic Publicity Department.*

2. If the direction of an impending force is known, stability can be increased by moving the line of gravity as close as possible to the edge of the base where the force is expected. A football lineman who expects to be shoved backward will lean forward toward the opponent and will spread his feet front to rear to widen his support base relative to the expected force.

3. Mass is directly proportional to stability. Other things being equal, the heavier person is the more stable.

4. The lower the center of gravity, the higher the stability. Crouching low will increase stability; lying prone will maximize it, particularly if the arms and legs are well spread out to increase the base and centralize the line of gravity.

Stability in sports situations is quite unlike the stability of solid objects. When a wrestler is being pushed backward, he can make a variety of physical adjustments to respond to the force. If his line of gravity is being shoved beyond his base of support, he can relocate his base by shifting his feet. He can push back or maneuver to redirect his opponent's force. He can also trick his opponent into thinking he will be pushing when in fact he intends to pull. Many of the martial arts emphasize this concept of taking advantage of an opponent's errors in weight placement.

Dynamic Stability

Dynamic stability is balance during movement, and it is much more difficult to identify the components of good dynamic stability than it is to describe the bases of static stability. It frequently happens that the line of gravity of an athlete will fall outside the base of support for a moment. As an example, in a sprint start the body weight is well ahead of the supporting foot, but before the body can fall forward the other foot moves ahead to provide support and the process repeats itself. Under conditions of rapid acceleration during linear motion, the line of gravity can be ahead of the supporting foot. In much the same way, the line of gravity will be outside the base of the feet during quick changes of direction.

During rapid directional changes, an athlete must lean inward to provide a turning moment that will offset the centripetal force acting at the feet. Balance will be lost if the coefficient of friction between the athlete's shoes and the ground is insufficient to provide the centripetal force, particularly during sharp turns at high speed. Balance will also be lost if the lean is not suitable to the radius or the turning speed.

Related to both dynamic and static stability are physiological factors such as kinesthetic sense, coordination, and inner-ear balance mechanisms. Experience, familiarity with the surface conditions, and quality of equipment are all involved in dynamic balance. Ice-skate blades should be kept sharpened, worn-out shoes should be replaced, and playing surfaces should be free of dirt or moisture. The experienced athlete makes allowances for poor conditions that might cause loss of balance.

SUMMARY AND DISCUSSION

Just about any movement a human makes is the result of the angular movement of a body segment. Angular motion exists when a body rotates around some axis, as for example when the arm flexes at the elbow joint. To achieve angular motion, an eccentric force must be applied. Such a moment of force, or torque, is the product of the unbalanced force acting and the perpendicular distance of the line of this force to the axis of rotation, the unit being the pound-foot (lb•ft) or newton-meter (N•m).

A moment may be internal, as when a muscle pulls on a bone to move it around a joint, or it may be externally applied to the body by gravity, friction, or some other force or resistance.

Angular measurements require special units. Displacement can be stated in terms of revolutions, degrees, or radians. Angular velocity is measured by dividing any of the displacement units by the time. Angular acceleration is found by dividing the change in the angular velocity by the time during which the change takes place.

Any point along the radius of a rotating body has both an angular velocity and a linear velocity. The latter is that tangential velocity with which the point would leave the body if it were released, and it is determined by multiplying the point's distance from the axis and its angular velocity. This concept has practical application in any sport where an object is to be thrown or struck with maximum speed, as is the case in pitching, batting, tennis serving, and discus throwing.

Angular and linear inertia are both measures of a body's resistance to a change in motion. Linear inertia is directly proportional to the mass of the body, but angular inertia (the moment of inertia) is related not so much to the mass as a whole as to *how* that mass is *distributed* around a particular axis. Thus, before you can judge the moment of inertia of a person or an object, you must know the axis around which the change in rotation will occur and how far most of the mass is located from that axis. Have you ever tried a novelty run keeping the legs straight at all times? It is fun and can be done with some speed, but fatigue soon sets in because the muscles have to work against the high moment of inertia of the fully extended legs. The leg normally flexes at the hip and knee during recovery, and this reduces the moment of inertia by bringing the leg mass closer to the hip joint, thereby requiring less effort from the hip flexor muscles.

Newton's three laws of linear motion have their angular equivalents. A rotating body has an angular momentum that will never change unless some outside torque acts. When such a torque does act, the rotating body will apply an equal and opposite torque, and if this is insufficient, the body will accelerate or decelerate around its axis. The angular momentum is the product of the body's moment of inertia and its angular velocity.

Since momentum will be conserved if there is no outside torque, a reduction of moment of inertia will produce an increase in angular velocity, and an increase in the moment of inertia necessitates a decrease in the angular velocity. Divers, gymnasts, and ice skaters utilize this law of conservation of angular momentum to speed up or slow down their rotations in the air or on the ice.

Whether a person wishes to do a vertical jump and make a full turn in the air or wants to complete some form of somersault, he or she must initiate this rotation while still in contact with the ground or board. To jump straight up and then decide to twist or somersault indicates a lack of understanding of the third law of rotational motion and the law of conservation of angular momentum. Such actions will cause opposite reactions, and the stunt will fail.

Anatomical levers are nearly all of the third class of levers. Most movements of the limbs require a lot of muscle force relative to the weight of what is being moved, but the force needs to be applied over a short distance compared with the distance the end of the lever moves. This is why we are able to move our arms and legs as rapidly as we do and over such a wide range.

Static stability is simply stationary balance, a position that tends to effectively resist most attempts to upset a person or object. A wide base or stance, a large mass, and a low center of gravity are indicators of stability.

Dynamic stability is balance in motion. The running, dodging, turning athlete who stays upright even when tripped is said to have great balance. Those factors that make for good static stability are of little value in dynamic situations, where a wide base and a low center of gravity are not ordinarily used in the game situation. More important are the principles of friction and centripetal force, along with physiological balance mechanisms.

Problems for the Student

1. Given a 10-lb force and a force arm of 1.5 ft, what is the torque?
2. Express in both radians and revolutions the angular displacement of a body that rotates 820 degrees.
3. A wheel rotates 240 times per minute. What is the angular displacement in radians after 10 s?
4. What is the angular velocity of a skater who is spinning 6 revolutions every 2 s?
5. Given a linear velocity for a point on a rotating radius of 27 ft/s and a radius length of 3 ft, determine the angular velocity.
6. Describe the most stable position that a wrestler might assume to avoid being pinned.

SUGGESTED READINGS

BUECHE, F. *Principles of Physics* (New York: McGraw-Hill Book Company, 1972).

HAY, JAMES "Moment of Inertia of the Human Body," in *Kinesiology IV* (Washington: American Association for Health, Physical Education, and Recreation, 1974), pp. 43–52.

KANE, T. R. and M. P. SCHER, "A Dynamical Explanation of the Falling Cat Phenomenon," *International Journal of Solids and Structures,* vol. 5, July 1969, pp. 39–49.

SATSCHI, W. R., J. DUBOIS, and C. OMOTO, "Moments of Inertia and Centers of Gravity of the Living Human Body," AMRL-TDR-63-36 (Wright Patterson Air Force Base, May 1963).

WIDULE, C. J. "Segmental Moment of Inertia Scaling Procedures," *Research Quarterly,* vol. 47, no. 2, May 1976, pp. 143–47.

CHAPTER 6

Work and energy

WORK

When an unbalanced force is applied to an object and causes it to move, work is done. Work, simply defined, is the overcoming of a resistance to produce a change in motion. It can also be described as the product of the displacement of the point of force application and the component of force acting in the direction of the displacement. From these definitions it should be apparent that if no motion results from a force application, no mechanical work is done, which is to say that there is no unbalanced force. Thus, work is done in an isotonic exercise but not in an isometric one, and work is done whenever an object is accelerated or decelerated by a force.

$$W = F \cos \theta \, d \qquad \textbf{(6–1)}$$

where W = work

F = unbalanced force magnitude

θ = angle between force vector and displacement vector

d = displacement of point of force application

The unit of work in the British system is the foot-pound (ft•lb), which can be defined as the work done when a one-pound force acts through a distance of one foot. In the SI system, the unit for work is the *joule* (J), which is the work done when a one-newton force acts through a distance of one meter. The term *joule* then could be replaced by the expression *newton-meter.*

Work is done in lifting a barbell vertically from the floor to any height. The force vector is in the same direction as the displacement vector, and so there is no angle θ to consider. The cosine of 0° is 1.0, and for all practical purposes, therefore, Eq. (6–1) could be expressed simply as $W = Fd$ for all problems involving vertical lifting.

When an object is lifted as a result of concentric contractions, the work done is positive. Negative work results from the lowering of the object by means of eccentric muscle action, because the force is being exerted in a direction opposite that of the displacement.

If the barbell in Figure 6–1 is lifted 6 ft overhead, 600 ft•lb of work will be done. If the athlete walks to the other side of the room while holding the weight overhead, he is doing no further mechanical work, because there is no motion of the weight in the direction of a force. The only force being applied is vertical, but the motion is horizontal. For the same reason, no work is done when an object is moving horizontally over a surface at a constant velocity, because the applied force is exactly offset by the opposing frictional force which means there is no unbalanced force.

Given the information that a 100-lb crate has been accelerated across

Figure 6–1 To lift a 100-lb barbell 4 ft vertically requires 400 ft•lb of work.

some surface over a 4-ft distance, it cannot be concluded that 400 ft•lb of work was done, because nothing has been said about the magnitude of force required to accomplish the motion. Nothing is known of the resisting frictional force or the coefficient of friction between crate and surface.

Now suppose there is a pulling force acting upon a crate by means of an attached rope at some angle to the horizontal (see Figure 6–2). To determine the effective force, that is, the force in the direction of the crate's displacement, the angle must be taken into account, as in Eq. (6–1).

Example 6–1. Given the data in Figure 6–2, calculate the mechanical work done.

Use Eq. (6–1):

$$W = F \cos \theta \, d$$

The 10-lb pulling force represents the net force available after the frictional force is subtracted from the applied force, but this is not the amount of force effective horizontally.

$$W = 10 \text{ lb} \times .866 \times 5 \text{ ft} = 43.3 \text{ ft•lb}$$

When an unbalanced torque acts through some angular displacement, *angular work* is done. This is the same definition given earlier for work but now stated in angular terminology. Angular work is done in initiating rotational motion, as in throwing a Frisbee, and in stopping such motion, as in catching a spinning ball.

Figure 6–2 A crate being displaced 5 ft by an unbalanced force of 10 lb acting at an angle of 30° from the direction of displacement.

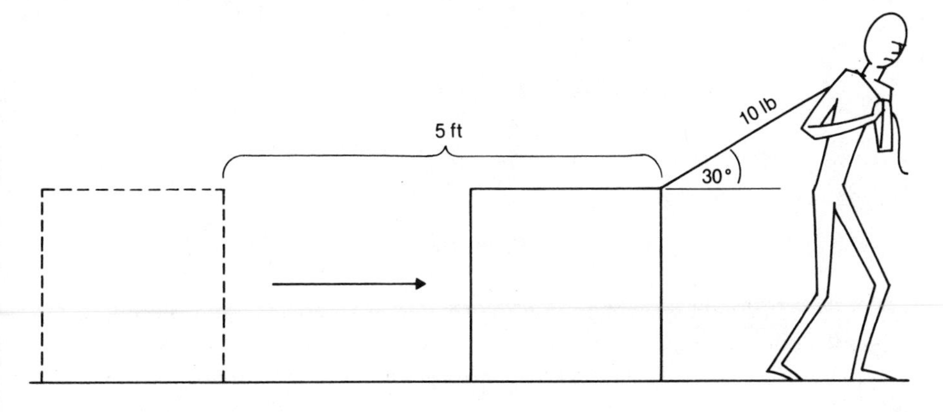

ENERGY

In mechanics, energy is the ability or capacity to do work, and it is classified as either kinetic or potential. There are a number of other forms of energy besides mechanical, such as nuclear, chemical, light, electrical, and thermal.

Kinetic Energy

Kinetic energy is that classification of energy that is associated with motion. A charging fullback, a flying javelin, or a swinging hammer all can do work as a result of their motion when they make contact with another object. Each possesses a force-producing capability.

The equation for kinetic energy is

$$\mathrm{KE} = \tfrac{1}{2} m v^2 \qquad \textbf{(6–2)}$$

where m = mass of moving body

v = velocity of body

The unit for kinetic energy is the same as that for work, foot-pounds (ft•lb) or joules (J).

Since work is just another name for energy that is transferred to an object by an unbalanced force, Eq. (6–2) can also be written

$$Fd = \tfrac{1}{2} m v^2 \qquad \textbf{(6–3)}$$

This equation can be derived by substituting in Eq. (4–1), $F = ma$, the value of a from Eq. (4–7),

$$a = \frac{v^2}{2d}$$

to get

$$F = \frac{mv^2}{2d}$$

$$Fd = \tfrac{1}{2} m v^2$$

It should be evident that the change in kinetic energy resulting from work done is directly proportional to the moving body's mass and to the square of its velocity. The positive work done on a body to give it its speed, then, is the kinetic energy of the body. When a moving body is stopped or slowed down, it gives up this amount of energy as a function of the negative work required to stop or slow it.

Example 6–2. What is the kinetic energy of a 200-lb halfback running at 24 ft/s?

$$\begin{aligned} \text{KE} &= \tfrac{1}{2}mv^2 \\ &= \frac{200 \text{ lb} \times (24 \text{ ft/s})^2}{32 \text{ ft/s}^2 \times 2} = 1800 \text{ ft}\bullet\text{lb (about 2441 J)} \end{aligned}$$

Example 6–3. Over what distance must a 100-lb force be applied to give an object a kinetic energy of 850 ft•lb?

$$\begin{aligned} Fd &= \tfrac{1}{2}mv^2 \\ 100 \text{ lb} \times d &= 850 \text{ ft}\bullet\text{lb} \\ d &= \frac{850 \text{ ft}\bullet\text{lb}}{100 \text{ lb}} = 8.5 \text{ ft} \end{aligned}$$

Potential Energy

Potential energy is energy a body has by virtue of its elevated position. Positive work is done to give a body the potential to do work, and when this potential is finally realized, the stored energy then becomes kinetic energy. As an example, a man climbing a ladder to a diving platform must do mechanical work. As long as he remains on the platform, he has a gravitational potential energy that is equal to the work he has done to get up there. When he at last dives from the platform he starts to develop kinetic energy while losing an equal amount of potential energy. Upon hitting the water he no longer has potential energy, and his kinetic energy is at its maximum.

Potential energy is the product of the weight and the height of displacement. Since weight is equal to mass times gravity,

$$\text{PE} = mgh \tag{6–4}$$

$$\begin{aligned} \text{where } mg &= \text{weight} \\ h &= \text{vertical displacement} \end{aligned}$$

In Eq. (6–4), height represents the difference in elevation from some arbitrarily chosen reference level. It may be ground level, but it might also be the level of an object's center of gravity. As with kinetic energy, the unit for potential energy is the foot-pound or the joule.

Conservation of Energy

The law of conservation of energy states that the energy within any closed system will remain constant. That is to say, energy cannot be created or destroyed. However, energy may be transformed from one form to another, as it is when heat or sound is produced in a collision of two objects. Kinetic energy is lost in the collision, but some other energy form is increased by the same amount. As indicated by Hopper, the diver who climbs the ladder loses internal energy as he does the work of climbing and produces heat energy.[1] Thus the chemical energy used results in work done, which becomes potential energy, after which kinetic energy is gained until work is finally done on the water, resulting in heat and waves. Energy has been conserved.

To fully reflect the law of conservation of energy, Eqs. (6–3) and (6–4) have to be combined to show that work done in a system is equal to the total of kinetic and potential energy. Therefore,

$$Fd\cos(90^\circ - \theta)d = \tfrac{1}{2}mv^2 + mgh\sin\theta \qquad (6\text{–}5)$$

Note here that Eq. (6–4) has been slightly altered to measure only the vertical height to which a body is raised. For that reason, it is necessary to multiply the displacement by the sine of the angle formed between the horizontal and the vertical. When the work is done horizontally, potential energy is not a consideration, and when the work is vertical, the sine of 90° is 1.0. So our concern is only with any angle between 0° and 90°.

To expand a bit on Eq. (6–5), consider a 70-lb wagon that has to be moved to a point 7 ft higher than where it now rests. To the destination there is an inclined pathway 20.5 ft long at a 20° angle to the horizontal. The alternative is to simply lift the wagon to the new place (see Figure 6–3). It should

Figure 6–3 A 70-lb wagon lifted to a platform 7 ft. high requires 490 ft•lb of work to be done. While it takes less effort to push the wagon up the incline, the same work is done.

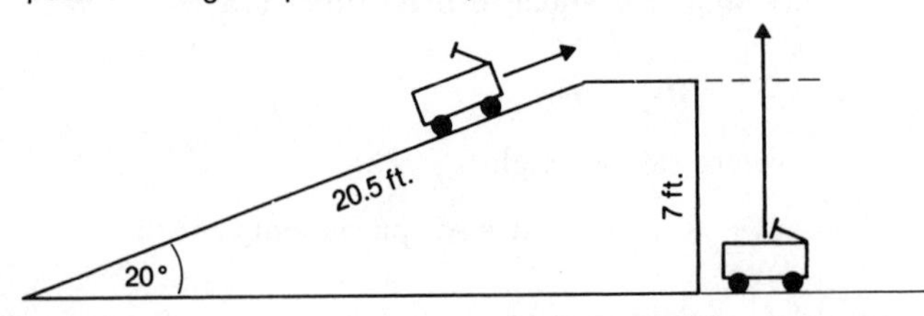

[1] B. J. Hopper, *The Mechanics of Human Movement* (New York: American Elsevier Publishing Company, Inc., 1973), p. 122.

be obvious that, either way, 490 ft•lb of work must be done. We cannot determine the kinetic energy, because there is no information about velocity. The potential energy can be measured, and since the potential energy is equal to the work done,

$$Fd \cos (90^\circ - \theta) = mgh \sin \theta \tag{6-6}$$

Hence, to lift the wagon vertically,

$$70 \text{ lb} \times 7 \text{ ft} \times \cos 0^\circ = 70 \text{ lb} \times 7 \text{ ft} \times \sin 90^\circ$$

$$490 \text{ ft•lb} = 490 \text{ ft•lb}$$

It is important to note that on the left side of the equation, which represents the work done, there is no angle between the force vector and the displacement vector and so the angle is zero degrees, for which the cosine is 1.0. On the potential energy side of the equation, the vertical lift is 90° from the horizontal and the sine is 1.0.

To push the wagon up the 20° incline over a distance of 20.5 ft,

$$70 \text{ lb} \times 20.5 \text{ ft} \times \cos 70^\circ = 70 \text{ lb} \times 20.5 \text{ ft} \times \sin 20^\circ$$

$$70 \text{ lb} \times 20.5 \text{ ft} \times .342 = 70 \text{ lb} \times 20.5 \text{ ft} \times .342$$

$$490 \text{ ft•lb} = 490 \text{ ft•lb}$$

The reason why 70° appears on the left side of the equation is again that that is the angle between the force and displacement vectors. If frictional opposition is ignored, about 24 lb of force will be needed to accomplish the task up the incline, but it has to be applied for 20.5 ft. Force is traded for distance.

The relationship between kinetic and potential energy can be illustrated by the raising of a weight to a particular height and the dropping of the weight back to its original level. Refer to Table 6–1.

Example 6–4. Consider a 3-lb block that has been lifted vertically to a height of 6 ft with 18 ft•lb of work thereby being done, so that the block has 18 ft•lb of potential energy. The block is dropped from the 6-ft height. What will be the kinetic and potential energies after the block has fallen 2 ft, 4 ft, and 6 ft?

Use Eq. (4–21),

$$v_y^2 = 2gs$$

to obtain the velocities, and Eq. (6–2),

$$KE = \tfrac{1}{2}mv^2$$

to get the kinetic energies.

After 2 ft,

$$v^2 = (2)(32\ \text{ft/s}^2)(2\ \text{ft}) = 128\ (\text{ft/s})^2$$

$$m = 0.09375\ \text{slugs}$$

$$KE = (\tfrac{1}{2})(0.09375\ \text{slugs})(128(\text{ft/s})^2) = 6\ \text{ft}\bullet\text{lb}$$

$$PE = 18\ \text{ft}\bullet\text{lb} - 6\ \text{ft}\bullet\text{lb} = 12\ \text{ft}\bullet\text{lb}$$

After 4 ft,

$$v^2 = (2)(32\ \text{ft/s}^2)(4\ \text{ft}) = 256\ (\text{ft/s})^2$$

$$KE = (\tfrac{1}{2})(0.09375\ \text{slugs})(256(\text{ft/s})^2) = 12\ \text{ft}\bullet\text{lb}$$

$$PE = 18\ \text{ft}\bullet\text{lb} - 12\ \text{ft}\bullet\text{lb} = 6\ \text{ft}\bullet\text{lb}$$

After 6 ft,

$$v^2 = (2)(32\ \text{ft/s}^2)(6\ \text{ft}) = 384(\text{ft/s})^2$$

$$KE = (\tfrac{1}{2})(0.09375\ \text{slugs})(384(\text{ft/s})^2) = 18\ \text{ft}\bullet\text{lb}$$

$$PE = 18\ \text{ft}\bullet\text{lb} - 18\ \text{ft}\bullet\text{lb} = 0$$

The reader should note that these energies can be computed more simply by using Fd. For example, after the block falls 2 ft, 3 lb × 2 ft = 6 ft•lb of kinetic energy. Since it is still elevated 4 ft, its potential energy is 3 lb × 4 ft = 12 ft•lb.

The example states that if the potential energy before a fall is equal to the kinetic energy at the end of the fall, the sum of the two is also constant throughout the descent. As the block gains kinetic energy during its fall, it loses an equivalent amount of potential energy.

Concepts about energy have the most meaning in sports in which an athlete is raised or some object is lifted. The horizontal bar performer puts energy into a system by muscular contractions that give his body potential energy, with which he can put his body into swinging movement. Thereafter, he manipulates his body to achieve maximum potential energy at the top of his swings and maximum kinetic energy at the bottom of his swings. The pole vaulter develops kinetic energy by speeding down the runway, and as he

TABLE 6-1
Relationships between Kinetic and Potential Energy for a 3-lb Block Lifted to a Height of 6 ft and Released.

Height	*PE*	*KE*	*Total Energy*
6 ft	18 ft•lb	0 ft•lb	18 ft•lb
5 ft	15 ft•lb	3 ft•lb	18 ft•lb
4 ft	12 ft•lb	6 ft•lb	18 ft•lb
3 ft	9 ft•lb	9 ft•lb	18 ft•lb
2 ft	6 ft•lb	12 ft•lb	18 ft•lb
1 ft	3 ft•lb	15 ft•lb	18 ft•lb
0 ft	0 ft•lb	18 ft•lb	18 ft•lb

plants and bends his pole, the pole receives and stores some of this energy. The vaulter must time his movements to take advantage of the quick release of the energy as the pole straightens. Upon clearing the bar, the vaulter himself now has potential energy by virtue of his elevated position, and this will change to kinetic energy as he falls to the pit. The conditions of his fall and landing are important applications of our energy equations.

INJURY PREVENTION

Based largely upon Eq. (6–3), $Fd = \frac{1}{2}mv^2$, some of the following principles directly apply to the study of injuries resulting from falls and collisions:

1. The greater the mass, the more severe the impact. As is evident in Eq. (6–3), the big person takes a harder fall than the small person.

2. The greater the falling velocity or impact velocity, the harder the fall or impact, because kinetic energy is proportional to the square of the velocity. There are limited situations in which the velocity of a fall is under the control of the coach, notably in pole vaulting and high-jumping, where if the pit is raised from ground level the falling distance and hence the falling velocity are reduced.

3. The greater the distance or the time over which the body decelerates upon impact, the smaller the average force that will be felt. When our vaulter lands, his kinetic energy is dissipated over some distance by the foam rubber in the pit. Compare this gradual deceleration over three or four feet to the near sudden stop experienced by vaulters landing in the sand or sawdust pits at ground level that were common years ago. Similarly, rubber mats, synthetic playing surfaces, and padded baseball gloves all serve the function of absorbing kinetic energy over a distance. Most athletes are

taught, or learn, to fall in an energy-absorbing manner by breaking the fall with a systematic, practiced maneuver involving joint flexions under the control of eccentrically contracting muscles, or by rolling upon impact, or by giving with the blow or the force being received.

4. The greater the area of impact, the less the force acting on any given point. This relates to the earlier discussion of pressure in Chapter 3. An athlete who lands on a broad surface of the body is less likely to be injured than one who lands on, say, an elbow or shoulder. As stated previously, this does not apply to diving, where a narrow entry is decidedly less painful.

Air bags in automobiles are an application of three of the principles mentioned: They provide a large contact area to reduce pressure; they reduce the inertial velocity of the rider upon impact by coming out to meet the rider's forward-moving upper body; and they allow for a gradual reduction of the kinetic energy of the rider.

5. For skidding or sliding types of falls, the abrasiveness of the landing surface must be considered. Friction between skin and a rough surface creates a great deal of heat and causes skin burns and abrasions. Ice skaters who fall do not suffer similar injuries, because of the reduced friction and the longer distances of sliding. The reader might note that this is simply an expansion of principle 3 above.

It can be concluded from these principles that the prevention of most injuries involves the reduction of impact force by means of an alteration of some component of Eq. (6–3) or Eq. (6–5). The impulse equation, Eq. (4–33), may also be used if the time of contact is known or sought. There is a saying that it is not the fall that hurts but rather the sudden stop. People have survived falls from great heights by being lucky enough to have landed in some yielding medium, such as mud, water, treetops, or brush.

By whatever means possible, d in the equation should be made as large as possible and v should be reduced. Young children seem to naturally flex at the hips and knees when they land from a jump, and, it is hoped, so do most adults. This increase in energy-absorbing distance can be readily appreciated by anyone who has ever tried jumping from a very low height and landing with rigid legs, the result of which is that kinetic energy is absorbed rather quickly and the landing shock is transmitted upward through the whole body, even jarring the teeth.

Example 6–5. A 150-lb pole-vaulter clears a height and falls 15 ft to a ground-level pit. What is the average force of impact if the pit yields a distance of 1 ft?

$$Fd = \tfrac{1}{2}mv^2 + mgh$$

$$v^2 = 2gs = (2)(32\text{ ft/s}^2)(15\text{ ft})$$

$$F \times 1 \text{ ft} = \frac{150 \text{ lb} \times (31 \text{ ft/s})^2}{2 \times 32 \text{ ft/s}^2} + 150 \text{ lb} \times 1 \text{ ft}$$

$$F = \frac{(2252.3 + 150) \text{ ft}\bullet\text{lb}}{1 \text{ ft}} = 2402.3 \text{ lb}$$

What would be the effect of using a raised foam rubber pit whose top is four feet above ground level and which yields 3 ft upon impact?

$$Fd = \tfrac{1}{2}mv^2 + mgh$$

$$v^2 = (2)(32 \text{ ft/s}^2)(11 \text{ ft})$$

$$F \times 3 \text{ ft} = \frac{150 \text{ lb} \times (26.53 \text{ ft/s})^2}{2 \times 32 \text{ ft/s}^2} + 150 \text{ lb} \times 3 \text{ ft}$$

$$F = \frac{(1650 + 450) \text{ ft}\bullet\text{lb}}{3 \text{ ft}} = 700 \text{ lb}$$

The falling velocities above were determined from Eq. (4–21) with falling heights of 15 ft and 11 ft. Any internal yielding of body tissue is of course ignored. Finally, the 2402.3-lb and 700-lb impact forces are distributed over a large area of the body and so the pressure is not necessarily great. The example's sole purpose is to convey the role played by distance and velocity in falls.

POWER

In the discussion of work, there was no mention made of a time factor. Work was simply the applied force times the displacement. But some people can work more rapidly than others, mechanically speaking, and can lift their own weight or an external object faster than can other people. The rate of doing work is the definition of power; therefore

$$\text{Power} = \frac{Fd}{t} \tag{6-7}$$

The unit of power is the foot-pound per second or the joule per second.

$$1 \text{ J/s} = 1 \text{ watt (W)}$$

$$1 \text{ ft}\bullet\text{lb/s} = 1.356 \text{ W}$$

$$550 \text{ ft}\bullet\text{lb/s} = 1 \text{ horsepower (hp)}$$

$$1 \text{ hp} = 746 \text{ W}$$

Power is important in explosive types of movements, such as vertical jumping, punching, batting, punting, shot-putting, and weight lifting. Power can be increased either by increasing the muscle force being applied in a particular skill or by reducing the time over which the muscles act to accomplish the task.

Example 6–6. Compare the power of two 154-lb men who each climb a vertical 14-ft ladder, one in 4 s and the other in 5 s.

$$\text{Power} = \frac{154 \text{ lb} \times 14 \text{ ft}}{4 \text{ s}} = 539 \text{ ft}\bullet\text{lb/s}$$

$$\text{Power} = \frac{154 \text{ lb} \times 14 \text{ ft}}{5 \text{ s}} = 431.2 \text{ ft}\bullet\text{lb/s}$$

The faster man uses 0.98 hp, compared with 0.78 hp for the slower man.

COLLISIONS

When two or more bodies collide, some energy is retained, some is transferred from body to body, and some is changed into another form of energy such as heat. There are three types of collisions, of which only two are found in sports situations:

1. *Perfectly elastic collisions* are those in which the relative velocities of the bodies before impact are exactly equal to their relative velocities after impact. No energy is lost or changed to other forms of energy. This condition never exists in sports, but it is most closely approached by the behavior of billiard balls striking one another.

2. *Imperfectly elastic collisions* are those in which the colliding bodies become temporarily deformed and then bounce apart or become separated from one another. Sometimes only one body is noticeably deformed, as when a basketball is being dribbled, a baseball or golf ball is being struck, or a barbell is bouncing off a rubber mat after being dropped. Occasionally, both colliding bodies are distorted, as when a tennis ball flattens while the racket strings become stretched or when a boxing glove and a face both yield during a blow. In all of these cases, one or both objects have their shape changed momentarily, usually causing some energy loss as heat, and then return to their original shapes during separation.

3. *Perfectly inelastic collisions* are those in which the colliding bodies do not separate after impact. Examples of inelastic collisions are an arrow hitting a target, soft clay hitting a wall, a bullet striking and entering a block of wood, and a classic football tackle (shown in Figure 6–4).

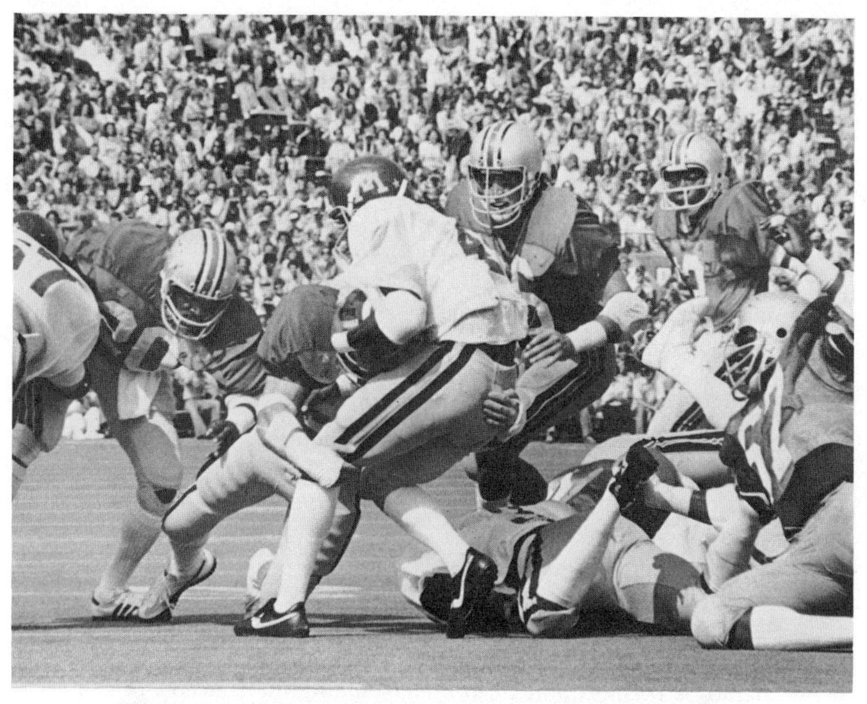

Figure 6-4 A football tackle may be considered an inelastic collision when the tackler hits and holds onto the ball carrier. *Photograph courtesy of the Ohio State University Athletic Publicity Department.*

ELASTICITY

Elasticity, or *restitution,* is the degree to which and the speed with which an object will return to its original shape after being temporarily deformed by some external force. However, the amount of deformity or the speed of resuming shape is not a determinant of rebound height. A steel ball dropped onto a steel surface will bounce higher than will a rubber ball. Imperfectly elastic collisions are the most common type in sports, and when they involve balls, we find one of the following conditions:

1. The ball moves against a fixed surface, as is the case when a tennis ball is struck against a backboard.
2. The ball is stationary up to impact while the striking surface moves, as happens when a golf ball is struck.
3. Both the ball and the striking surface move toward one another, as in baseball batting or tennis stroking.

THE LAW OF CONSERVATION OF LINEAR MOMENTUM

Everything discussed so far regarding energy and a good many of the linear motion principles join together to form the basis for the law of conservation of linear momentum. This law is really the modern equivalent of Newton's third law. It states that the total momentum of any isolated system of bodies is unaffected by any actions that occur between the different members of the system, which is to say that the total momentum of two bodies before they collide will be equal to the total momentum after the impact, if perfect elasticity is assumed. But of course some energy is always converted, and so kinetic energy is changed. It should be easy to see the direct relationship to the third law, in which F_1 due to body 2 is equal to F_2 due to body 1.

Any loss of momentum experienced by one body in the system is balanced by an equal gain of momentum by the other body after an elastic collision, so that the combined momentum is unchanged. The time of impact will always be the same for both bodies. Conservation of linear momentum can be expressed by the equation

$$m_1v_1 + m_2v_2 \text{ before impact} = m_1v_1 + m_2v_2 \text{ after impact}$$

But since this is true only for perfectly elastic collisions, its value to biomechanics is perhaps limited to understanding the concepts of conservation of energy and the third law. If the masses and velocities of two bodies are known before impact, the final velocities after impact cannot be determined unless the coefficient of elasticity between the bodies is known.

It should now be possible to see the interrelationships between conservation of energy, conservation of linear momentum, Newton's laws, and impulse. Equations (4–1), (4–34), and (6–5) are all related, and it must be repeated that kinetic energy is lost to varying degrees in any collision. The more nearly perfectly elastic the collision, the less the energy loss to other forms. What we call *elasticity* is simply the activity of gaining or losing kinetic or potential energy. The force that causes an elastic deformation of a ball is called a *stress,* and the deformation itself is the *strain,* which is the effect of the stress. A stress is a ratio of applied force to the area over which the force acts, and it may stretch, compress, twist, or shear a body.

COEFFICIENT OF ELASTICITY

In all collisions, the nature of the materials making up the involved bodies must be considered when the effects of the collision are evaluated. According to Chao, impulse forces also depend upon the shapes and sizes of the colliding bodies, as well as upon their impact geometry.[2] For any pair of bodies striking one another, there is a coefficient of elasticity or restitution, which is the ratio of the relative velocities of the two bodies before and after colliding. A coefficient of 1.0 represents a perfectly elastic collision and indicates that no kinetic energy is lost. A coefficient of 0 represents a perfectly inelastic collision, in which the bodies adhere after impact.

To find the coefficient of elasticity of a ball dropped from a height of 6 ft that bounces up to a height of 4 ft, use the equation

$$e = \sqrt{\frac{h_2}{h_1}} \tag{6-8}$$

where e = coefficient symbol

h_2 = height to which the ball bounces

h_1 = height from which the ball was dropped

$$e = \sqrt{\frac{4}{6}} = \sqrt{.6666} = .816$$

A correctly inflated basketball will have a higher coefficient than one that is underinflated. A handball will bounce higher than will a squash ball when both are dropped from the same height onto the same surface. A bean-bag or a lump of clay will not rebound much on any surface. Temperature plays a role in elasticity, in that a ball generally will bounce somewhat higher when it has been heated than when not.

Example 6–7. A steel ball has a coefficient of elasticity of .9 on a particular surface. If it is dropped from a height of 10 ft, how high will it bounce?

$$.9 = \sqrt{\frac{h_2}{10 \text{ ft}}}$$

$$h_2 = 10 \text{ ft} \times .9^2 = 8.1 \text{ ft}$$

The coefficient of elasticity can also be expressed as

$$e = \frac{\text{velocity after impact}}{\text{velocity before impact}} \tag{6-9}$$

The similarity between Eqs. (6–9) and (6–8) should be evident if reference is

[2] E. Y. Chao and others, ''Mechanics of Ice Hockey Injuries,'' in J. L. Bleustein, ed., *Mechanics and Sports* (New York: American Society of Mechanical Engineers, 1973), p. 150.

made to Eq. (4–21), $v_y^2 = 2gs$, because for vertically falling balls, the velocity is a function of the height.

When both the striking surface and the ball move toward each other, Eq. (6–9) is changed to

$$e = \frac{v_a - v_b \text{ after impact}}{v_b - v_a \text{ before impact}} \tag{6–10}$$

When a pitched baseball is struck center to center by a bat, the speed of the ball after impact will depend upon the following:

1. The coefficient of restitution of the bat and ball
2. The mass of the bat and its velocity before the hit
3. The mass of the ball and its velocity before the hit

It has been stated by Daish that in batting and golf, the velocity of the bat or club is of greater significance in achieving ball velocity than is the mass of the implement being used.[3] He indicates that the tightness of the grip at impact is of little consequence, but this is contradicted by Hatze, who has found that in tennis, while a firm grip increases the impulse, it also increases the vibrational shocks that have to be absorbed by the hand.[4]

Most momentum and collision problems, such as those that follow, are necessarily contrived. They try to depict actual sports situations, but many assumptions must be made before they can have any meaning in analysis. For instance, when a hockey player crashes into the boards, it is not possible to measure the coefficient of elasticity between the boards and the player's body and so there is no way to predict the bounce after the impact. But contrived or not, some examples might help the student to a better concept of the principles involved.

Example 6–8. Consider a 224-lb fullback carrying the ball through the line at a velocity of 20 ft/s. He is tackled head-on by a 192-lb linebacker. If the tackler manages to stop the fullback cold in his tracks, what is the linebacker's velocity at the time of the tackle?

Since the total momentum after impact is zero, it follows that the total momentum of the system before impact has to be zero as well. Thus,

$$m_1 v_1 + (-m_2 v_2) = 0$$

[3] C. B. Daish, *The Physics of Ball Games* (New York: Sterling Publishing Co., 1972), p. 20.

[4] Herbert Hatze, "Forces and Duration of Impact, and Grip Tightness during the Tennis Stroke," *Medicine and Science in Sports,* vol. 8, no. 2, 1976, pp. 88–95.

Therefore,

$$m_1 v_1 = m_2 v_2$$

$$224 \text{ lb} \times 20 \text{ ft/s} = 192 \text{ lb} \times v_2$$

$$\text{Linebacker's velocity} = 23.3 \text{ ft/s}$$

(Note that it is not necessary to convert weight into slugs in this problem.)

Example 6–9. The players in Example 6–8 collide again. This time the fullback jumps over the line at 10 ft/s and is met head-on in midair by the linebacker, who is traveling at 20 ft/s. The tackler holds on tightly and the two players fall to the turf. Determine the horizontal velocity $v_{1,2}$ of the players after this inelastic collision and the direction in which they move before hitting the ground.

$$m_1 v_1 + (-m_2 v_2) = m_{1+2} v_{1,2}$$

$$224 \text{ lb} \times 10 \text{ ft/s} + (-192 \text{ lb} \times 20 \text{ ft/s}) = (224 \text{ lb} + 192 \text{ lb}) v_{1,2}$$

$$-1600 \text{ ft/s} = 416 v_{1,2}$$

$$v = -3.8 \text{ ft/s}$$

The negative sign indicates that the pair are moving in the same direction as that of the linebacker's original motion.

Example 6–10. A 90-lb girl steps out of a canoe toward a dock and is moving at 4 ft/s. The 60-lb canoe naturally reacts in the opposite direction. What will be the canoe's velocity?

$$90 \text{ lb} \times 4 \text{ ft/s} = (60 \text{ lb}) v$$

$$v = 6 \text{ ft/s}$$

and so, of course, the girl gets wet.

REBOUND

Closely related to the study of collisions and elasticity is the subject of rebound, which is a factor in most sports that involve balls. One of the biggest problems facing a beginner in learning such sports as tennis, racquetball, and squash is judging how the ball will bounce off the court floor, wall, or ceiling, depending on the game. Even experienced players in one racket sport must go through a period of adjustment to the different elasticity coefficient and rebound they encounter in a new sport.

As a ball approaches a stationary surface for an oblique elastic collision, its flight line forms an *angle of incidence* with a line drawn perpendicular to the surface. As it then rebounds, the exit flight line forms an *angle of reflection* with the same perpendicular line. Figure 6–5*a* illustrates these angles for a nonspinning ball bouncing off the floor, while Figure 6–5*b* shows that in a bat-ball collision, the perpendicular line is drawn from a tangent at the point of contact between the bat and ball surfaces. In this figure, the baseball is being popped up.

For perfectly elastic bodies in collision, the angle of incidence equals the angle of reflection, and all the energy is conserved. The height of rebound equals the height from which the ball is dropped. In sports there are no perfectly elastic collisions, and so the angle of reflection will always be somewhat greater than the angle of incidence as measured from the perpendicular line, how much greater depending, of course, on the coefficient of elasticity. The difference in rebound angle or height is a measure of the energy lost to other forms.

A teacher or coach will probably find it easier to use the floor as a reference line for the angles of incidence and reflection, because it makes more sense to students to say that a ball takes a *lower bounce* because it has lower elasticity than it does to say that it has less rebound from a perpendicular line that cannot be seen.

The rebound behavior of a ball on a surface is affected by a number of factors, including the coefficient of elasticity (which in turn depends on the ball's inflation, age, and temperature) and the ball's velocity, friction, and spin rate. A ball with topspin generally will take a lower bounce relative to the

Figure 6–5 *(a)* A non-spinning ball rebounding off a floor surface. *(b)* A nonspinning ball being struck by a swung bat. In both figures, the angles of incidence and reflection are drawn relative to a perpendicular line.

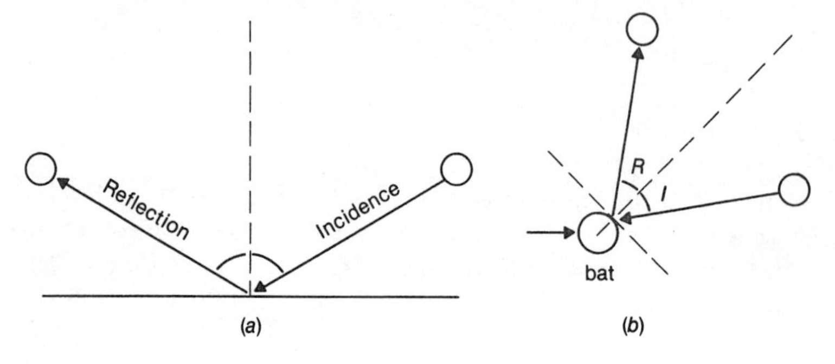

ground than a nonspinning ball, and a ball with backspin will take a higher bounce. In topspin situations, the part of the ball contacting the court is moving backward and there is a forwardly directed frictional reaction to this force. The result is a forward acceleration, which contributes to the lower rebound angle. A topspin serve in tennis imparts a high linear and angular velocity to the ball, which causes the ball to curve sharply downward and thus creates a high angle of incidence relative to the ground. This may partly account for the apparently higher bounce taken by such serves when most other balls would take a lower bounce. The interrelationships between the ball, the surface, the elasticity coefficient, the velocity, the angular spin, friction, and the incidence angles offer a complex challenge to researchers.

Surface friction can act as an eccentric force to initiate rotation in a nonspinning ball when it makes floor contact. The ball has a horizontal velocity component, which is opposed by friction to provide a moment of force and hence some topspin. This also occurs in the delivery of a bowling ball, which first slides and then rolls.

SUMMARY AND DISCUSSION

Mechanical work is done when the point of application of a force moves some distance in lifting, lowering, pushing, or pulling an object. Work gives the object kinetic or potential energy, the former being the energy of motion and the latter being the energy of position. The unit for work or energy is the foot-pound or the joule.

The law of conservation of energy indicates that energy cannot be created or destroyed, but some forms of energy can be converted to other forms. A body receives kinetic energy or potential energy, and an equal amount of work must be done to absorb this energy when the body stops moving. The average force required to absorb a moving body's energy is inversely proportional to the distance that the force can act. A safety net under a circus trapeze will give several feet when an acrobat falls into it. Injury prevention is based on providing this kind of yielding surface as well as on reducing the falling velocity and increasing the area of contact to reduce pressure.

Anatomically, some of the shock of a fall or landing is taken by special structures. The arch of the foot provides resiliency when the landing is on the feet. Similarly, the knee cartilages (menisci) and the intervertebral discs absorb some shock. Landing on heavily fleshed parts of the body offers some cushioning against impact.

Collisions, rebound, and elasticity are interrelated and are parts of the law of conservation of linear momentum, which states that in any closed system of colliding bodies no momentum is lost. This assumes a perfectly elastic collision, in which all the energy is retained. An imperfectly elastic collision is the type most frequently encountered in sports, and a perfectly inelastic collision is less common. One or both colliding bodies temporarily deform and, if they are elastic, will quickly reform, as when a basketball bounces off the floor or a tennis ball rebounds from the racket strings. The closer the basketball bounces to the height from which it is dropped, the higher the coefficient of elasticity is between it and the floor. Naturally, if it bounces up higher, it must also have a higher velocity leaving the floor than a ball whose coefficient with the floor is lower.

The rebound of a ball depends upon its elasticity, its speed, its incoming angle (angle of incidence), its spin, and the type of floor it will bounce on.

Problems for the Student

1. What force is required to change the momentum of a body from 20 kg•m/s to 80 kg•m/s in a period of 3 s?
2. What work is done when a 30-lb unbalanced force acts to move a body 6 ft?
3. What work is done when a 40-N unbalanced force acts over a distance of 12 m?
4. A 75-g wooden block is struck by a 15-g bullet, which becomes embedded in the block. If the bullet has been traveling at 300 m/s at impact, what is the velocity at which the block will move after the impact?
5. Consider a 130-lb man on a trampoline who bounces up 12 ft from the bed. What force is exerted to bring the trampolinist to a halt if the bed yields 3 ft when he lands on it?
6. How fast must a 170-lb halfback run to gain the momentum necessary to stop a 220-lb fullback running 25 ft/s?
7. If the coefficient of elasticity between a ball and a surface is .77, how high will the ball bounce when dropped from a height of 8 ft?

SUGGESTED READINGS

BISHOP, P. J., "Dynamic Response Criteria for Ice Hockey Helmet Design," in P. V. Komi, ed., *Biomechanics V–B* (Baltimore: University Park Press, 1976), pp. 299–305.

BROER, M. R., and R. F. ZERNICKE, *Efficiency of Human Movement* (Philadelphia: W. B. Saunders Company, 1979).

CAVAGNA, G. A., F. B. SAILENE, and R. MARGARIA, "Mechanical Work in Running," *Journal of Applied Physiology,* vol. 19, 1964, p. 249.

Cochran, A., and J. Stobbs, *The Search for the Perfect Swing* (Philadelphia: J. B. Lippincott Company, 1968).

Fenn, W. O., "Work against Gravity and Work Due to Velocity Changes in Running," *American Journal of Physiology,* vol. 93, 1930, p. 433.

Gurdjian, E. S., and H. R. Lissner, "Mechanism of Concussion," in F. G. Evans, ed., *Biomechanical Studies of the Musculo-Skeletal System* (Springfield, Ill.: Charles C Thomas, Publisher, 1961) pp. 192–208.

Kovacic, C. R., "Impact-Absorbing Qualities of Football Helmets," *Research Quarterly,* vol. 36, December 1965, p. 420.

Mellen, W. R., "Superball Rebound Projectiles," *American Journal of Physics,* vol. 36, 1968, p. 845.

CHAPTER 7

Analysis of movement

It is one thing for a student to learn the laws of physics as they seem to apply to human movement and quite another to make use of this acquired knowledge. How does one proceed to analyze movement in the gymnasium or on the playing field? What types of analysis are there, and what are some methods that can be used by teachers and coaches?

All coaches and teachers of physical education are concerned, at one time or another, with movement analysis and some of the more practical applications of biomechanics. Occasionally the focus is on observing a good performance so as to determine what makes it good. More often, the teacher's concern is with poor execution and the means by which faults may be corrected.

The average physical education major is not likely to do laboratory research in biomechanics, at least not without a great deal of further study. However, it might be reasonable to expect that prospective physical-educators or coaches will utilize a number of kinesiological and biomechanical principles that they recognize as relevant to their sports interests.

Biomechanical information can be applied in the planning of safer sports and activity programs, particularly by taking into account the principles of energy and collisions. It also may be employed in doing some basic qualitative analyses of sports skills. It should enable the student to comprehend analytical articles and studies. And it can provide a basis for speculation about better ways to perform skills. Even if no improvements actually result, the very process of thinking about a sport in a fresh manner is likely to promote a better appreciation of the mechanics that apply and the relationships of principles to activity.

TYPES OF ANALYSIS

There can be a wide range of goals and sophistication levels in the application of biomechanical principles to sports problems. Whether one is simply using a mechanical term in teaching a skill to a beginner or is doing research in a laboratory, the common purpose is to learn about or improve human movement. A rough classification of approaches to and purposes of biomechanical analysis may be useful.

Kinesiological Analysis

A kinesiological analysis seeks to identify the joints, muscles, and bone levers used in a skill as well as the sequence and degree of their involvement. This kind of information might come from published electromyographic studies that have been done on the skill; it might be found in kinesiology textbooks; or it might come from a coach's experience and familiarity with anatomy as it relates to his or her sport. It is difficult for a student with a limited background in kinesiology to make correct anatomic analyses of motion, because muscle involvement is not always what would appear to be logical. However, to the extent that an anatomic analysis can be done with moderate accuracy, it is a necessary first step in the prescription of exercises needed to develop muscles for a particular sport or to improve an athlete's joint flexibility.

Biomechanical Analysis

Biomechanical analysis has already been defined and can range from simple to complex, depending upon the nature of the problem, the purposes of the analysis, and the capabilities of the analyst. It may help an athlete cut performance time by a fraction. It may serve to point out means by which an inch can be added to a performance. Or it may simply be used by teachers or coaches to broaden understandings of sports skills. There are three categories of mechanical analysis:

1. A *qualitative* analysis includes visual and photographic observations, which usually result in a description or a judgment of the good and the weak points of a given performance. This is the approach most commonly

used by teachers and coaches, but it is very often a technique evaluation rather than a biomechanical one. Visual analysis has the obvious advantage of not requiring expensive equipment but suffers from limited accuracy and is most effectively practiced by an expert coach with an experienced eye. One of the great frustrations for the new coach or teacher is the inability to pinpoint the causes of poor performance by a student. Without instant replay, the teacher must depend upon the senses to be able to quickly see what took place. With the luxury of film or videotape and the time to view repeatedly a single performance, the chances for correctly diagnosing an error are enhanced.

The filming process itself is very critical, and most ordinary game films are not of much use in analysis of an individual, because of the probability of poor camera angle, background, or light. The time and expense needed to photograph individual performers and then to study the film for perhaps several hours is usually justified only when the subject is an outstanding athlete for whom every small advantage is worth the time and effort. The faults of lesser athletes are generally those of technique or are due to physical limitations, and an observant coach can recognize these without a lengthy analysis.

How does technique analysis differ from biomechanical analysis? Technique can be defined as the sequence of actions an athlete follows in performing a sports skill. In other words, it is a specific manner of execution. Biomechanics permits the explanation of and the rationale for what occurs, both the good and the bad, during each movement of a performance. While all aspects of technique are within the potential control of the performer, many aspects of mechanics are not, and these must be understood because they will inevitably affect the outcome. But in many sports, it is not enough to possess good technique based upon sound mechanical laws. In these sports, some of which are individual and some team, the opponent presents constantly varying situations to which the athlete must adjust. Qualitative analysis becomes difficult when the performer is being affected by such factors as adaptation to the game conditions, competitive pressures, and the running out of playing time. These varying conditions are somewhat less a factor in such sports as high-jumping, shot-putting, billiards, bowling, and archery, where the athlete has a degree of control over the playing conditions.

2. A *quantitative* analysis involves the measurement and recording of hard data about movement and goes well beyond qualitative analysis because of its emphasis on instrumentation.

This type of scientific analysis may be done on any of several levels, ranging from research that has immediate applicability to sports, all the way to research that deals with narrow problems of seemingly little practical consequence but may nevertheless contribute to the overall body of knowledge about human movement. It is possible for a laboratory theorist who knows little about a specific sport to raise valid questions and even to postulate techniques that might not occur to an expert in the sport who has been indoctrinated in a particular style. Scientists in the lab are aided by interesting and very technical measuring and recording devices, including high-speed cameras, motion analyzers, force platforms, and computers (see Figure 7–1).

Funded research directed toward the improvement or development of items of sports equipment such as football helmets or running shoes may provide findings that have obvious potential value to players and to the manufacturers who finance this type of research. These investigations may be done in university laboratories or in the research and development departments of manufacturers.

However, interest in human movement is not limited to coaches, teachers, athletes, and manufacturers. Biomedical and aerospace engineers

Figure 7–1 Modern technology aids researchers from various fields in the study of human gait. While the subject in the photograph steps over a force plate, a television camera feeds a computer with images from the lights attached to his shoulder, hip, knee, ankle, and toe. *Photograph courtesy of the Ohio State University Medical Photography Department.*

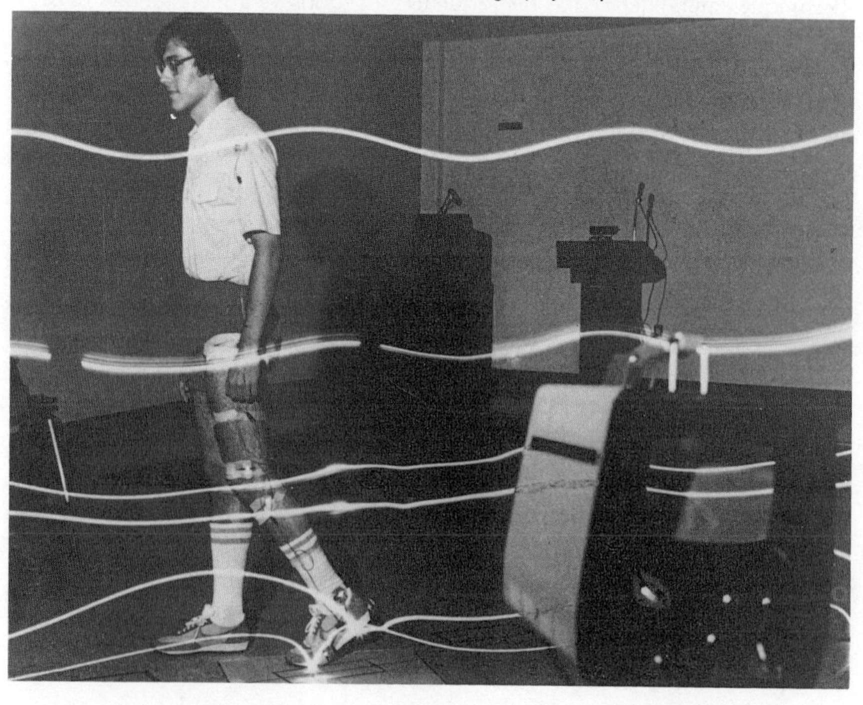

have been involved for years in studying locomotion, prosthetic devices, and movement under restricted circumstances. Many major university hospitals have gait laboratories devoted to the study of normal and abnormal gait.

3. *Biomechanical speculation,* while not in itself a method of analysis, must be based upon an understanding of mechanics and anatomy and may be employed by anyone who is expert in a sport and wonders about alternative ways that a skill might be performed. This sort of thinking may be necessitated by a rules change that suggests a need to change or adapt the current technique or by the introduction of new equipment, as happened when the fiber glass pole was brought out some years ago. In either case, some thought by an expert in the sport who has biomechanical training may hasten the adaptation to the new rule or equipment.

In contemplating how to improve performance in your own sport specialty, ask yourself questions such as the following:

> *What major result is desired?* For example, in volleyball the blocker must be able to jump high. What can be done to help the player gain height? We can certainly try to increase leg strength, which is the traditional route to follow, but are there some mechanical principles to be reviewed and applied?
>
> *Is the present stance or technique the most effective, and has anyone ever attempted to change it?*
>
> *Do the rules or the aesthetics of the sport specifically dictate the technique now in use?*
>
> *If some mechanical principle is being ignored or incorrectly applied and a new technique is hypothesized, is the proposed change within the rules?* Is it strategically feasible? Will athletes be physically capable of making the suggested new movement? Will this change negatively affect performance in some other aspect of the sport?
>
> *How must technique change to get the best results from the use of newly developed equipment?*

THE ANALYZERS

While there is a great deal of interest in human movement on the part of engineers in such fields as aerospace and biomedicine, it is presumed that sports biomechanics is largely the concern of coaches, physical-educators, and sports scientists. These professionals deal on a day-to-day basis with young people in all grade levels in the schools and colleges.

Physical Education Teachers

Analysis by physical education teachers is, almost by necessity, of a qualitative nature. The teacher is not as concerned as the coach with the concepts of farther, faster, and higher or with game scores and meet records. Nevertheless, the same mechanical principles that affect the Olympic performer also affect the third-grade pupil who is just learning a skill. Teachers must be familiar with these principles because, even when the teacher is not in a position to do individual analyses of students, he or she can effectively use mechanical knowledge to select safe and purposeful activities and calisthenic exercises for the physical education program.

Teachers in the gymnasium must often make instant qualitative assessments of student performance, and their ability to do accurate analyses depends upon a number of factors including their expertise in the particular activity, their understanding of the mechanics involved, and their background in kinesiology. It is soon evident to the novice teacher that showing pupils *how* to execute a movement is only the beginning of the teaching process. Making corrections is considerably more difficult than introducing skills.

After seeing that a student is performing incorrectly, what is the next step a teacher must take? The cause of the error might be found in the act itself, or it might be traced to some flaw in such preparatory phases as the approach, the windup, and the backswing. The broader the skill background of the analyst and the higher the level of personal expertise, the more likely that a quick and accurate evaluation of the problem can be made.

But no teacher is skilled in all sports and stunts, and it is not uncommon for teachers from elementary grades on through college level occasionally to have to teach activities in which they have had no experience. Many sports have common movement patterns, and a preliminary biomechanical study of an unfamiliar activity might well reveal some significant mechanical principles that underlie another, more familiar sports skill. These principles then could suggest to the teacher some logical and safe teaching methods and progressions. For instance, teachers who have a baseball background will be better able to teach tennis if they see the similarities of movement patterns in the two sports. Certainly the tennis serve is not unlike an overhand pitch, and these skills share such common principles as accounting for the summation of forces, developing the highest possible linear velocity at the end of the lever, and anticipating the tangential path that will be taken by an object released from a circular path.

In summary, one of the neglected curricular areas in the professional preparation program is instruction in how to see errors, trace their causes, and make corrections. For the practicing teacher, the chief obstacle is the short time available and the large number of students in a class. Individual help cannot always be given.

Coaches

Analysis by athletic coaches is also usually qualitative except insofar as times, distances, heights, or scores are recorded. When coaches analyze the performance of one of their athletes, they are typically using as a criterion some style employed by a number of top performers. This process of looking for deviations from or conformity to a common style is, strictly speaking, technique analysis rather than biomechanical analysis. Technique analysis is both justifiable and necessary, because a coach is expected to be aware of currently effective styles or strategies, and it is of paramount importance for a coach to teach a proven skill technique that produces good results. The coach's job is often dependent on the season's record.

At most levels, then, coaches do not make true biomechanical analyses of athletic performances, which is not to say that they cannot or should not do so. As is the case with physical education teachers, the average coach does not have an abundance of time or equipment for analysis. Unlike teachers, many coaches, especially of football and basketball, do have the use of motion picture cameras and projectors and perhaps videotape recorders. While such equipment is seldom of scientific grade, some basic analysis can be done.

Aside from the question of the coach's time, there is the issue of priorities. For younger athletes, improvement is very often a direct result of motivation, both intrinsic and extrinsic, to work harder and longer. The developing athlete also improves as a function of maturity, size, strength, coordination, and inherited talents, but these factors may also constitute limitations on an athlete's development. Given the physical capabilities, the avenue to success is in technique improvement, which is where a good coach can add not just inches but feet to novice performance, can cut seconds from times, and can add points to scores.

The statement has been heard many times that it can be unwise to imitate champions because we know that some have achieved that status in spite of imperfect techniques. This cautious concept has been overplayed, because the top athletes, for the most part, have form that *is* worthy of imitation and indeed are, have been, and always will be imitated. Biomechanicians frequently lose sight of the fact that the vast majority of young athletes will never become champions and that most coaches will never have the opportunity to work with world-class performers. Yet most athletes *do* strive for excellence at their own level and deserve the same support and attention that seems often to be reserved by the theorists only for the champion.

For the average coach, no understanding of sports biomechanics is as necessary as a fundamental knowledge of the common techniques of a sport and an ability to observe the technical flaws of pupils. Not all coaches are equally perceptive about selecting the right one of several possible styles to suit the special characteristics of an individual athlete or about placing each team member in the most appropriate event or position. This might require the matching of the mechanical requirements of a particular technique to the physical capabilities or limitations of the athlete. A given style may be totally unsuited to the build or maturity of a particular athlete. Especially when training young children in sports, it is important for coaches not to impose on them the styles used by adults. Even for mature athletes in, say, track, a coach should be prepared to teach one beginning high-jumper the Fosbury style and another the straddle style. It is this kind of judgment ability that accounts for the success of some coaches.

While all techniques can certainly be *explained* biomechanically, they have seldom evolved through someone's conscious application of biomechanical knowledge. It is probably safe to say that most successful athletes and coaches have achieved their good results without the benefit of biomechanical understanding at other than a common-sense or intuitive level. This is an unfortunate circumstance that will no doubt change as the science of sports biomechanics continues to advance. Evidence of the possible gains that may be expected is already coming out of intensive scientific programs in Eastern Europe.

Having just made a somewhat negative case for the value of biomechanics in coaching, we now should state positively that all coaches and physical-educators ought to have a good working knowledge of biomechanics, for the obvious reason that it makes them more competent professionals. It is not necessary that a coach know the mechanical basis of each and every sport, but each one should know the underlying biomechanical principles of his or her own sport. Familiarity with the principles and language of biomechanics is very necessary for a full understanding of the technical literature available in every sport.

Laboratory Research

Analysis by laboratory researchers is nearly always quantitative. The subject under study may be an athlete, but biomechanical inquiry is not limited to human movement. Athletic equipment is examined in the laboratory to discover improvements that might serve to improve protection to the athlete, to improve playing surfaces, to improve flight characteristics of balls and other throwing objects, to measure the capabilities of vaulting poles or swimming-lane lines, to determine means by which friction can be increased or decreased, and so on.

Such investigations are not necessarily the special province of sports biomechanicians. A person need not know much about football in order to study or compare football helmets. It is not necessary to be a hockey player to measure the impact-dissipating ability of a face mask. Obviously, a sports background is helpful in these types of studies, but what is common to all biomechanicians is the physics, engineering, and mathematics training they have undergone.

The communication of the meaning of research outcomes to the teacher and coach is an ever-increasing concern of biomechanical researchers. They have recognized the existence of a gap between theory and practice, but often a research study by itself is only one piece of a large puzzle that will have full value only when completed. The combined contributions of workers in biomechanics, exercise physiology, and motor learning will someday have a tremendous impact on sports teaching and training.

METHODS OF ANALYSIS

There are basically two approaches to analysis of human movement, and each can emphasize either anatomic or mechanical aspects of motion or both. Whichever method, visual or instrumental, is used, the first assumption to be made is that the teacher, coach, or researcher is personally skilled in the sport being studied. It would be very difficult for an inexperienced person to analyze a skill of any complexity. The second assumption is that the person doing the analyzing has had or is taking a course or two in anatomic kinesiology and sports biomechanics. The mere possession of knowledge of fundamental physics and gross anatomy is generally insufficient for the study of sports movement.

As a start in the analytic process, it would be logical to go through a number of kinesiology and biomechanics texts to determine whether satisfactory analyses of the sport in question are available. Next, a check through the indexes of periodicals devoted to the sport may reveal articles of value. Where these steps prove unproductive, the coach or teacher may then undertake an analysis, the depth and scope of which will naturally depend upon the available equipment and the capabilities of the analyst as well as upon the skill of the subject athlete.

Visual Methods

Visual methods are commonly employed by coaches and teachers. A performance is observed, errors are recognized, and corrections in style are made. It should be noted that even a successful performance may have had flaws that should not go uncorrected simply because the athlete has won. Might the athlete have done better?

Where possible, the human eye should be assisted by motion pictures, which can be viewed repeatedly, preferably on a projector that can show single frames as well as slow motion. Many skills, notably gymnastics and diving, involve highly complex movements done with such speed that even an experienced coach might have difficulty seeing possible mistakes without the aid of film.

For both anatomic and mechanical analyses, whether done visually or with instrumentation, it is helpful to adopt some format that organizes the examining process by breaking down a skill into its major components. One possible approach follows:

1. Preparatory or ready position: In most sports there is an early stance taken while the athlete awaits some signal or is contemplating the skill to be performed. Such a position thus may have no real effect on the skill, or it may be an integral part of the skill. The position may require stability, as for a back dive, may be best suited for quick movement, as in sprint starts, may enhance force production because it places muscles under stretch, may provide for greater distance over which force is applied, or may be arbitrary and unrelated to what will follow.

2. Approach: Such sports as the high jump, pole vault, long jump, and diving, to name but a few, involve an approach that is critical to the final performance. Very often a faulty execution can be traced to an approach error.

3. Skill execution: Depending upon the complexity of the skill, the normal sequence of movements and their purposes in the sequence must be understood by the analyst. Here, every movement has significance and must be carefully studied.

4. Termination: As with preparation, approach, and execution, the purpose and importance of movements made at the end of a performance must be clear to the observer. For instance, in diving and gymnastics it is necessary to finish with good form, whereas in pole vaulting the landing is not evaluated and only safety is of concern. In other sports, such as bowling, shot-putting, and discus throwing, the athlete must be careful not to step over the restraining line or ring.

In doing a visual analysis, the observer must watch the athlete at all stages: preparation, approach, execution, and termination. It may be the purpose of the analysis to detect errors in technique or mechanics, or the analysis may be an anatomic one intended to study the sequence and degree of involvement of muscles in a performance. In the latter case, it is important to note the role played throughout by gravity so as to judge whether the muscular contractions are concentric or eccentric in nature. Finally, the coach should keep in mind the possibility that faults in performance might be attributable to nonmechanical causes such as fear, anxiety, illness, or lack of strength.

Much can be overlooked when the visual approach is used, but it is a necessary method despite its shortcomings. Student teachers need supervised experience in observing performance critically, and this training may be expedited through the repeated viewing of loop films or videotapes.

Instrumental Methods

Instrumental methods are used to only a limited degree by teachers and coaches and usually at a comparatively simple level. The use of stopwatches and measuring tapes and the recording of points are quantitative. A track or swimming coach often measures not only the final race time but also the split times at critical points. These may tell more about a performance than does the total time. The angle of lean during a sprint start or of entry in a swim start can be measured from films. A measuring tape can be used to record horizontal or vertical jumping distances, and accuracy in archery is easily determined quantitatively. All of these are rather elementary methods for data collection, but these devices do have the virtues of portability, economy, practicality, and reasonable accuracy.

The next most practical tool is the high-speed motion picture camera, which may be considered the staple, indispensable, and most widely used instrument for analysis of human movement. Movies have long served as the primary data source in athletics, and as ever-faster films become available, much present-day research continues to employ cinematography.

The filming process itself must be done correctly; the ordinary game film is of little value. Plagenhoef provides excellent treatment of cinematography.[1] Typically the film is projected onto a screen or special viewing surface and measurements such as joint angles, displacements, and accelerations are taken directly from the screen. When done by hand, this type of data collecting is tedious and time-consuming. Motion analyzers are commercially available; these devices project the image onto screen grids, from which

[1] Stanley Plagenhoef, *Patterns of Human Motion* (Englewood Cliffs, N.J.: Prentice-Hall, Inc. 1971), pp. 7–16.

measurements can be taken more quickly and accurately, especially when the analyzer is linked to a computer. (See Figure 7–2.)

As is often the case with biomechanical measuring instruments, motion picture analysis is usually preceded by the identification of the segmental links to be measured and the determination of their masses, centers of gravity, and moments of inertia. These are available through various tables, such

Figure 7–2 A number of commercial motion analyzers are available. The NAC Model 160B Film Analyzer is shown. *Photograph courtesy of Instrumentation Marketing Corporation, Burbank, California.*

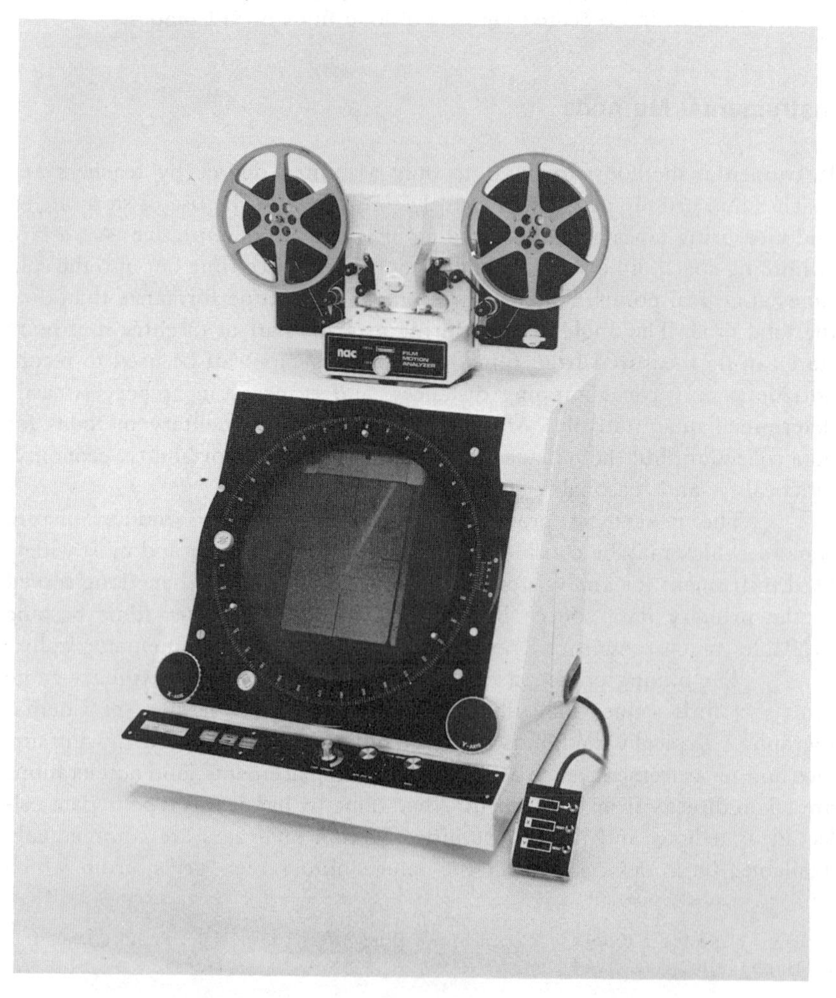

as those produced by Dempster.[2] The segmental mass centers may be marked directly on the subject to be filmed in a manner that will show up in the film and thus expedite analysis. Stick figures are commonly traced from evenly spaced frames by connecting the joint centers, and such drawings speed the measurement process by highlighting the essential aspects of the movement. It is of course important to know the camera speed in order to determine the time represented by each frame of the film.

A vast array of increasingly sophisticated instruments are becoming available for researchers, and many of these are well described by Miller and Nelson.[3] Whereas the movie camera has the virtues of being portable and unobtrusive to performance, most of the other devices have the disadvantage of being stationary or of restricting the subject because of their wiring or their need for a controlled environment. For those reasons, research tends to be confined to the laboratory, an unnatural athletic environment.

The equipment in current use includes force platforms, treadmills, wind tunnels, oscillographs, photoelectric cells, stroboscopes, electromyographs, electrogoniometers, motion analyzers, and computers. For some studies, telemetry has eliminated the need for some wiring of athlete to instrument.

The extent and timing of muscle contractions can be recorded by the electromyograph, and this devise is especially valuable when it is linked to other instruments, such as the force platform, which can measure horizontal and vertical force components exerted by an athlete standing on it. The electrogoniometer, or elgon, is used to record joint angles in two dimensions. Most of the equipment cited requires considerable training to use or for interpreting the data generated.

NONBIOMECHANICAL FACTORS IN PERFORMANCE

One cannot begin to explain all human movement in terms of only one discipline. It should be made clear that biomechanical factors play but a small part in sports performance and training even though mechanical laws are ever-present and not to be ignored.

The whole athlete must be considered before any judgment can be made about the causes of poor performance. In those few cases where an athlete succeeds despite a seeming violation of the currently accepted styles, it is necessary to examine the role being played by determination, willingness to train, reflexes, strength, tactical sense, confidence, courage, and concentra-

[2] W. T. Dempster, "Free Body Diagrams as an Approach to the Mechanics of Human Posture and Motion," F. G. Evans, ed., in *Biomechanical Studies of the Musculo-Skeletal System* (Springfield, Ill.: Charles C Thomas, Publisher, 1961).

[3] D. I. Miller and R. C. Nelson, *Biomechanics of Sport* (Philadelphia: Lea & Febiger, 1973).

tion. In examining the total performer it may be necessary to measure or speculate about any or all of the following nonbiomechanical factors:

1. The quality of the athlete's nervous system, vision, reflexes, reaction time, and intelligence
2. The athlete's psychological condition during practice and during competition
3. Physiological requirements of the sport such as strength, flexibility, and endurance
4. The state of the athlete's physical health when the skill is being learned as well as during a meet performance
5. The amount of facility required by the skill compared with that possessed by the athlete
6. The athlete's age and general readiness to learn the skill
7. The nature of the environment during the performance, including the importance attached to the event, the quality of the opposition, the size and makeup of the audience, and the presence of relatives and friends, scouts, or the media
8. The athlete's intrinsic motivation to learn and to succeed
9. Any unique intellectual, aesthetic, or artistic requirements of the sport
10. The sum effect of all of the performer's previous athletic experiences as they relate to the present sport, including the quality of the teaching and coaching that has been received, the nature of the equipment and facilities utilized, and the quality and quantity of competition faced to date

Obviously, some of the above factors are within the power of the coach or the athlete to control and some are not, but they must all be considered before any meaningful *total* analysis can be made or before any improvements in performance can reasonably be expected.

SUGGESTED READINGS

Ariel, G., "Computerized Biomechanical Analysis of Human Performance," *Athletic Journal,* vol. 54, no. 7, March 1974, p. 54.

Basmajian, J. V., *Muscles Alive: Their Functions Revealed by Electromyography* (Baltimore: Williams & Wilkins Company, 1974).

CAVANAGH, P. R., "Electromyography: Its Use and Misuse in Physical Education," *Journal of Health, Physical Education, and Recreation,* vol. 45, no. 5, May 1974, pp. 61–64.

CURETON, T. K., "Elementary Principles and Techniques of Cinematographic Analysis as Aids in Athletic Research," *Research Quarterly,* vol. 10, no. 2, 1939, pp. 3–24.

HOPPER, B. J., and J. E. KANE, "Analysis of Film: The Segmentation Method," in J. Wartenweiler, E. Jokl, and M. Hebbelinck, eds., *Biomechanics: Technique of Drawings of Movement and Movement Analysis* (Basel, Switzerland: S. Karger AG, 1968), pp. 42–44.

O'CONNELL, A. L., and E. B. GARDNER, "The Use of Electromyography in Kinesiological Research," *Research Quarterly,* vol. 34, May 1963, pp. 166–184.

PLAGENHOEF, S., "Methods for Obtaining Kinetic Data to Analyze Human Motion," *Research Quarterly,* vol. 37, March 1966, pp. 103–112.

ROBERTS, E. M., "An Introduction to Cinematography: Cinematography in Biomechanical Investigation," in J. M. Cooper, ed., *Selected Topics on Biomechanics: Proceedings of the C.I.C. Symposium on Biomechanics* (Chicago: The Athletic Institute, 1971), pp. 41–50.

SUKOP, J., K. L. PETAK, and R. C. NELSON, "An On-line Computer System for Recording Biomechanical Data," *Research Quarterly,* vol. 42, 1971, p. 101.

SUSANKA, P., "Computer Techniques in the Biomechanics of Sport," in R. C. Nelson and C. A. Morehouse, eds., *Biomechanics IV* (Baltimore: University Park Press, 1974), pp. 531–34.

CHAPTER 8

Applications to physical education and sports

There are far too many sports and physical education activities to attempt to cover all in one book, let alone in a single chapter. The activities chosen for treatment here are representative rather than inclusive and are either fundamental motor skills, common calisthenic exercises, or sports skills. The intent is not to give any activity a full mechanical or technique analysis but, rather, to apply as many of the principles of sports biomechanics as possible to some selected skills in order to illustrate the uses of the laws presented in the preceding chapters.

CALISTHENIC EXERCISES

Athletic coaches and physical education teachers typically use a range of exercises as part of an overall program to develop the strength, endurance, and flexibility of their pupils. The extent of the exercises' use depends to some degree upon the availability of various exercise and weight training machines, which offer systematized mechanical means to achieve fitness as well as the added motivation that is associated with gadgetry. Whether such equipment is used or the athlete does exercises without such aid, a number of mechanical considerations must be reviewed for their bearing on the safety and effectiveness of particular methods of exercise.

Knee Bends

Knee bends can be done in various ways, each with a particular outcome in mind. Moderate knee bends are an easy and effective means for developing the quadriceps femoris muscle group (rectus femoris, vastus medialis, vastus intermedius, and vastus lateralis), which extends the leg at the knee joint. The hamstring group (biceps femoris, semitendinosus, and semimembranosus) is also active in knee bends, as it extends the hip joint.

By eccentric contraction of the hip and knee extensors, the body weight is slowly lowered (negative work is done) and, in the process, the knee and hip joints move farther from the midline of the body. The result is that an external gravitational moment of force acts and must be opposed by the muscles. The deeper the knee bend, the greater the moment. In the rise back to an erect position, positive work is done by the same muscle groups, now contracting concentrically, and the moments due to gravity are diminished back to zero.

Pediatricians, orthopedists, and physical educators have, for the most part, taken an official position condemning the *deep* knee bend as an undesirable exercise because of the great stress placed on the knee joint. The widespread use of exercises such as the duck waddle done in the deep-bend position is a thing of the past. However, it can be debated whether deep knee bends should be considered dangerous to the joints of everyone. After all, the full-squat position is a common stance in many cultures. Baseball catchers hold this deep position for long periods, and weight lifters do knee bends deeply with certain lifts. But because there is no way for a teacher to know which pupils may be candidates for knee problems, routine prescription of deep knee bends for entire classes should be avoided.

Doing the exercise to even a half-squat is beneficial, and as illustrated in Figure 8–1, it lends itself nicely to enough modifications to permit anyone to experience success.

A range of difficulty levels is shown in the figure. If the knee bends are all done to the same squat depth, the easiest method is one in which the performer's line of gravity is as close as possible to the toes. This may be done by flexing the trunk and holding the arms in a forward elevated position and thus shifting the weight center forward to reduce the moment arm (Figure 8–1*a*). The short moment arm minimizes the torque acting around the knee joint (Figure 8–1*a*). This method might be used by unfit or older persons or by those who have had knee problems. Of course, there are other exercises to develop knee and hip extensors that do not require the handling of full body weight and can be prescribed in place of knee bends.

The next level of knee-bend difficulty requires that the trunk be kept in a nearly vertical position while the hands are on the hips (Figure 8–1*b*). In

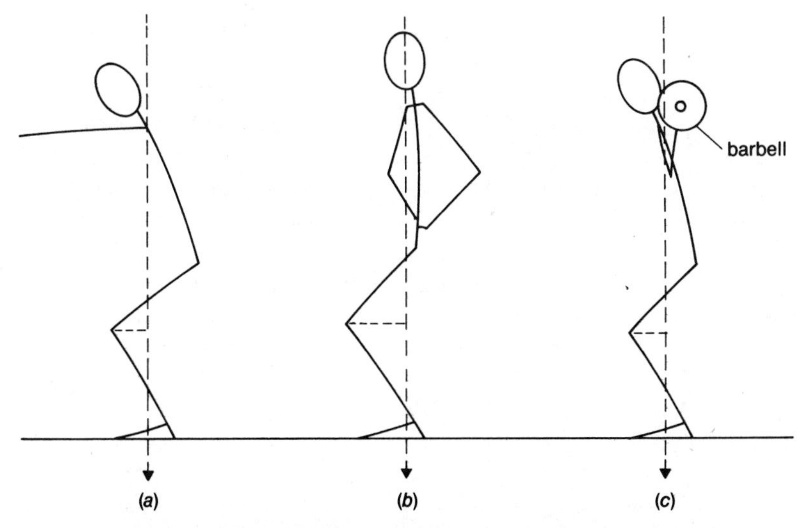

Figure 8-1 Three levels of difficulty in doing knee bends. At the left (*a*), the moment arm at the knee is shorter than in the second figure (*b*), and the torque is therefore less. The added weight on the right (*c*) increases the torque even though the moment arm appears to be the same as that of position (*a*).

this position, the line of gravity falls near the heels, thereby increasing the perpendicular distance from the knee joint to the line of force.

For those who desire more muscle development than might be possible from the two methods described, weight can be added (Figure 8–1*c*). Thus the external resisting moment of force is increased by the amount of weight being moved times the length of the moment arm. It should be noted that in all forms of knee bends, the moment arm continuously changes during the exercise, being zero in the erect position and maximum at the completion of the squat. Without adding external weights, knee bends can be made more difficult if they are done on only one leg at a time. Again, there are other, more demanding exercises for the knee and hip extensor muscles, but these require the pressing of weights with the legs.

In noncompetitive weight training, as distinct from the sport of weight *lifting,* the knee joint should be protected by the avoidance of full knee bends while heavy weights are borne on the shoulders. The strain felt at the knee increases as the depth of flexion increases. This is because the external moment of force becomes larger on account of the added weight and the gradual lengthening of the moment arm, which becomes greatest at full squat. Further, as knee flexion exceeds 90°, there is a potential shearing force acting on the protective ligaments of the joint, and so moderate knee bends are recommended for untrained persons working with weights.

The muscles used in this exercise can be strengthened or given more endurance in any of four ways. First, the weight can be held constant while the knee bends get progressively deeper over a period of several weeks. By this means, more and more work is done in raising and lowering the body from the deeper squats, and therefore the muscles must exert force over a longer distance. Second, the knee-bend depth can be held constant while weight is gradually increased. This overloading will create more torque and require more work, since a larger force must be applied over a given distance. Third, endurance can be improved by performing more repetitions of the exercise over a period of training while both depth and weight are held constant. Fourth, the rapidity with which a given number of knee bends are performed can be increased over a period of time. Here the work is constant but the power is improved.

Sit-ups

Sit-ups may be done in any of several ways and can be effectively moderated to achieve particular outcomes. There are two moments to consider. An *external* torque is measured as the product of the upper-body weight and the perpendicular distance from the line of gravity of the moving upper segment to the hip joint. The *internal* moment is found as the resultant of the products of the hip flexor muscle forces and their respective force arms. When the magnitude of the internal muscular moment exceeds that of the gravitational external moment, a sit-up is possible, and a concentric contraction of the hip and spine flexors lifts the trunk up to the sitting position. When the body is returning to the supine position, it is generally lowered under the control of the eccentrically contracting hip and spine flexors.

Performance targets include (1) the completion of a specific number of sit-ups, (2) a specific time to complete a specific number of sit-ups, or (3) a certain number of sit-ups to do in a particular time. Since the center of gravity of the upper segment is being raised, work is being done in sitting up and potential energy is being developed. Doing that work at a faster pace brings in a power dimension. While some strength is developed, muscular endurance is probably the major outcome of increasing the number of sit-ups. How many sit-ups are enough? What is to be gained when a person who now can do 100 sit-ups tries for 200? It is probable that there is one figure that can be set for a minimal fitness level and a much higher goal for athletes whose sport requires strength and endurance in the abdomen-hip region.

Aside from aesthetic considerations, a flat and firm abdomen is worth developing because it provides the only anterior protection for the lower body organs. It is particularly necessary in those contact sports where abdominal blows are frequent.

The best means by which to develop these muscles is still a matter of controversy. Although Clarke reports that there are some 52 different variations of the sit-up exercise, three specific methods are most commonly discussed.[1] They are the straight knee with rigid torso, the straight knee with curling torso, and the bent knee with curling torso. The questions raised center around which style best develops abdominal muscles over hip flexors and which style has the least adverse effect on the lumbar spine. The muscle at issue is the psoas major, which is a hip flexor with upper attachments on the lumbar spine that it might tend to hyperextend.

Electromyographic and X-ray studies of sit-ups indicate that there may be slight stress on lumbar discs, but no evidence appears anywhere that any student has received significant injury or postural deformity from having done sit-ups in any of the usual styles. Considering the numbers of other types of injuries sustained in sports, the controversy over injuries in sit-ups may be overdone.

As to which style most effectively develops the abdominal musculature (rectus abdominis, internal obliques, and external obliques), no method has achieved consistently better results than the other methods, and so it is left to the individual judgment of the teacher or coach, at least until new findings come along.

From a mechanical standpoint, regardless of the sit-up style, there are two external gravitational torques acting. With the hip joints and lower spine acting as the fulcrum, the upper body offers a resisting moment, and the legs on the other side of the fulcrum also have a moment acting upon them. When the sit-up is performed, the hip flexor muscles pull equally on both attachments, with the result that there is a tendency for the legs to rise at the same time that the torso rises. For this reason, some people are unable to perform a sit-up unless their feet are held down in some manner.

In a curling-type sit-up, the head is lifted at the beginning of the sit-up, and this overcomes the resting inertia of at least part of the body to be lifted. A sidelight of this head lift is that it tenses the rectus abdominis, which then stabilizes the pelvis and flattens the lower back against the floor. Consequently, those coaches and physical educators who are concerned about protecting students from possible lower-back discomfort or injury might insist on a curl type of sit-up rather than a straight-back type. In keeping with the most widely accepted style, the curl sit-up might be taught with the knees flexed, even though it is still being debated whether this in fact shortens the psoas major and prevents its strong contraction.

[1] "Exercise and the Abdominal Muscles," *Physical Fitness Research Digest,* H. Harrison Clarke, ed., series 6, no. 3, July 1976 (President's Council on Physical Fitness and Sports).

A common teaching error found in physical education situations and in athletic conditioning is the demand that an entire class or group perform sit-ups, or any other exercise, identically, say with hands clasped securely behind the head. Except where the students are being tested against some norm, there is no justification for such rigid conformity. Individual differences must be considered. One youngster may not be able to do a single sit-up when restricted to a specific style, while another might be able to do a hundred. Both are being denied the opportunity to develop. There is no "right" way to do sit-ups.

For the weaker student, inability to do sit-ups as prescribed could result in embarassment in class and a resulting dislike for exercise. How much better it would be to allow students to swing their arms (transfer momentum from part to whole) as an aid to sitting up until such time as their musculature develops sufficiently to enable them to perform in another manner. Another fairly easy method is to do the sit-up or curl-up with the arms at the sides. This brings the upper segment's mass center closer to the hips and thereby reduces the resisting external moment of force.

The stronger student can be given some added weight to hold. This increases the moment by adding to the gravitational force. The moment can also be adjusted by the manner of positioning the weight. If held on the chest, the weights do not increase the torque as much as they would if they were held behind the head. Either way, the overloading of the abdominal and hip muscles should produce an increase in strength of these muscles, whereas simply increasing the number of sit-ups performed would enhance endurance.

Sit-ups can be made more difficult through the use of an inclined board. Two mechanical factors involved here are (1) the increase in work necessitated by the higher lifting of the center of gravity and (2) the increased range of motion required to achieve a vertical sitting position (Figure 8–2).

Figure 8–2 In normal situps (left) the center of gravity of the torso is raised and work is done. On an inclined board, the center of gravity must be lifted to a higher level and the torso moves through a range exceeding 90°.

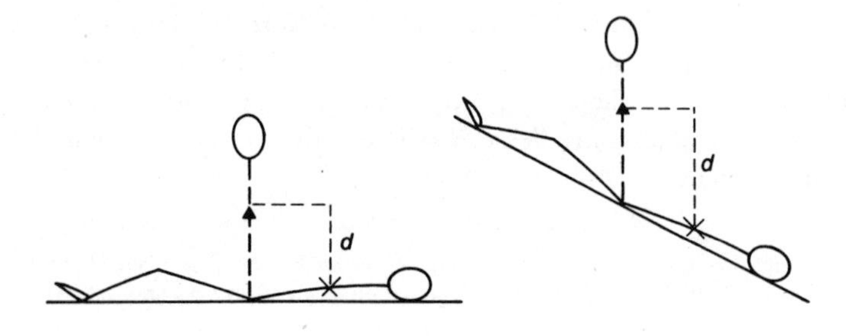

Leg Lifts

Leg lifts in one sense are the reverse of sit-ups in that the segment that is fixed in sit-ups is the segment that is moved in leg lifts. There is some similarity of muscles used, depending on the styles that are being compared. The double leg lift is neither a necessary nor a desirable exercise, but it is discussed here as another example of how an exercise can be modified to suit individual needs.

The double leg lift places an even greater strain on the lumbar spine than does the straight-leg sit-up because it usually involves lifting the feet a few inches from the floor and then holding the legs still for a period before lowering them. There is a fairly high gravitational moment acting and being opposed by the internal moment of the hip flexors and knee extensors. Again, the major expressed concern is for the effect of the psoas' pulling on the lumbar attachments.

A teacher who feels that this is an important exercise should at least modify the exercise according to the student's abilities. The weaker person should be permitted to flex the knees, thereby reducing the moment arm by bringing the mass center closer to the hip joint axis. The hip flexors will then be able to lift the legs more easily and the involvement of the knee extensors will be reduced (see Figure 8–3).

Some of the potential harmful effects of the leg lift can be eliminated if the students prop themselves up on their forearms and hold the trunk in a curled position during the leg lift. The inclined board may be used in leg lifts, the head being on the uphill side and the hands grasping the top of the board.

Pullups

The pullup is an excellent, safe exercise for developing the elbow flexors (biceps brachii, brachialis, and brachioradialis) and the shoulder extensors (principally the latissimus dorsi). Since the body is lifted and lowered, there is a concentric and eccentric contraction of the same muscles, and so there is ex-

Figure 8–3 Leg lifts can be made easier if the student is allowed to flex the knees (*a*). If there is concern for the adverse effects that double leg lifts might have at the lumbar spine, the exercise can be done with the torso curled and propped up on the elbows (*b*).

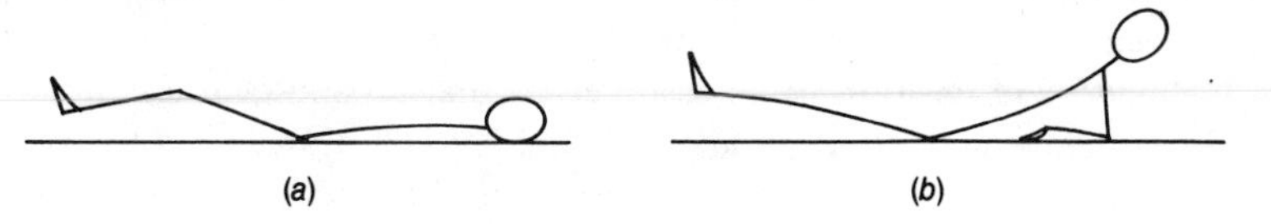

ercise benefit in both directions. Positive work is done as the body weight is raised, and negative work is done on the muscles as the body is lowered. The lowering should always be controlled, because there is enough potential energy in the up position that a sudden drop could cause joint or ligament damage.

There are some variations of the conventional pullup. One involves doing pullups on an inclined ladder but with the feet remaining on the floor so that the full body weight is never lifted. Another is the bent-arm hang in which the student may be timed while holding the static position with the chin just above the bar. Where a bar is not available, a climbing rope may be substituted.

The limits to the ability to do pullups are set by the weakest link in the system. A small percentage of students do not have the grip strength to do a simple extended-arm hang from a bar and so cannot be expected to do pullups until the finger flexors are developed.

The position of the hands on the bar is an anatomic consideration more than a mechanical one. With the hands in a supinated position, that is, with the palms toward the face, the biceps brachii is in a more favorable position than when the hands are pronated, because the distal attachment of the muscle on the radius is in a direct alignment in supination. In pronation the radius crosses the ulna and the biceps attachment rotates medially, causing the muscle tendon not to be in a direct line with the pull. The other two elbow flexors are not as greatly affected by the hand position, and the latissimus dorsi, being a shoulder extensor, is unaffected by hand position on the bar.

Push-ups

The push-up is one of the most commonly done exercises and presents opportunities for modification to satisfy individual needs. Like the pullup, it is a safe and effective exercise. It develops the elbow extensor muscles (triceps brachii, anconeus) and the shoulder horizontal flexors (pectoralis major, anterior deltoid). Because the body is maintained in rigid extension during push-ups, many other muscles are statically contracted to stabilize the ankles, knees, pelvis, and trunk.

There should be no insistence that an entire group of persons do push-ups in some uniform manner. For an elderly or debilitated individual, push-ups can be done vertically against a wall, a variation which requires very little strength. Another style, slightly harder, is to push against a table top, a secure bench, or a set of stairs, that is, against any object which allows an inclined body position to be assumed so that a good deal less than body weight is being moved. At about the same level of difficulty is the push-up done from the hands and knees, which shortens the resistance arm and thus the gravitational force. For stronger students, resistance to the push-up can be increased by adding weight to the performer or by elevating the feet to any level higher than the hands. Parallel-bar dips require the lifting and lowering of full body weight through elbow extension and flexion, but this exercise involves somewhat different shoulder muscle involvement from that of the normal push-up.

Any calisthenic exercise should be carefully chosen by teachers or coaches to fill specific needs of students. All too often a class is put through a series of exercises that have no intent except to serve vaguely as a warm-up. Exercises can contribute to strength, flexibility, or endurance, but the teacher must know which muscles are involved, which students need special prescriptions, and the number and frequency of repetitions that will accomplish the goals. Not only must the exercise be purposeful, but it must also be safe. The biomechanical considerations are the moments that act, the weight that is moved, and the vulnerability of particular weak joints. In physical education class, conformity of all to one style is valid only when tests are being given to measure present status or accomplishment against a norm. For selected groups of conditioned athletes, conformity may be demanded because of the esprit de corps that can be derived from working in unison.

The teacher who understands kinesiology and mechanics can write prescriptions for students that state how many times, how often, and how rapidly an exercise should be done as well as simply *how* to do it.

The key questions to be answered in doing an anatomic analysis are: Which joints are involved? What is the nature of the movement that occurs at each joint? Does the exercise lift or lower body weight against gravity? Which muscles concentrically contract to achieve the identified movements? There are no guides to how many times, how often, or how rapidly a given exercise should be performed at a particular age, for a particular sex, or at a particular level of fitness.

As regards flexibility exercises, the increased range of motion at a joint is specific to that joint. In other words, each joint is considered separately because there are no transfer effects. A person may be flexible in the spine and inflexible in the shoulders, or flexible in the right shoulder and not in the left. The range of motion at a joint is limited by the capsule and surrounding tissue and ligaments, by the tendons that cross the joint, and by the structure of the joint itself. There is nothing to be done about the last of these, but the soft tissue can be stretched over a period of time.

The two methods for increasing flexibility are the static and the ballistic. The former involves moving a body part at the joint to its comfortable limit, then going beyond the limit and holding the uncomfortable position for a few seconds before relaxing and repeating. This is a relatively safe and effective approach. Equally effective but more hazardous is the ballistic method, in which a body segment uses its momentum to cause stretching. As one example, if the arms are raised to a forward elevated position and then horizontally extended with some speed until they are stopped by the antagonistic tissue in front of the shoulder, there will be a stretching but at the risk of tearing or damage to the joint capsule. Because the arms are extended throughout, they have a high moment of inertia, which, when coupled with a high angular velocity, gives them a large angular momentum ($A = I\omega$). The sudden stopping of this angular momentum can be traumatic to whatever is subjected to the stretching. Therefore, this and similar exercises done with a bounce are to be generally avoided. Static stretching gets the same results without the attendant dangers.

FUNDAMENTAL MOTOR SKILLS

The movement-education trend that has been widely adopted at the elementary school level has brought with it renewed attention to many biomechanical principles. The concepts of force, space, time, and flow are derived from mechanics and are well applied to many of the fundamental skills. If one were to draw up a rather long list of what might be considered fundamental skills, not everyone would agree with its content. The activities that follow do not constitute a full list of fundamental motor skills but were chosen because they are common and lend themselves to application of mechanical principles.

Walking and Running

Walking and running are certainly fundamental. We learn to walk on the average by the age of one and start running soon after. The technical distinctions between these two forms of locomotion are very simple. As long as a person has at least one foot on the ground at all times, he or she is walking. Running begins as soon as both feet are allowed to be simultaneously off the ground. (Speed of movement and angle of lean play no role in distinguishing between a walk and a run; in fact it is quite possible for a person to choose to run more slowly than he or she walks.) Furthermore, whereas in walking there are times when both feet are in contact with the ground at once, in running there is never more than one foot on the ground at any one time.

There is a broad range of walking styles, some natural and some learned. The walk of a fashion model is quite different from a stiff military march. The double-time walk of a marching bandplayer has little in common with the studied walk down the aisle of a bride. The animated walk of a kindergartner is very unlike the labored gait of an octogenarian.

In analyzing a walking or running stride, defined here as the completion of one step with each foot, there are three phases to consider:

1. The previous stride ends and a new stride begins with some kind of foot strike. In walking and jogging this is a heel strike, and in sprinting it is the ball of the foot which hits first. From this initial contact until the body's center of gravity passes over that foot, there is a *restraint* phase. It is so called because, in walking, the heel strike is well in front of the line of gravity, and at contact there is a momentary rearward frictional reaction that restrains forward progress. In running, the foot strike is almost directly under the center of gravity, so that the restraint phase is either very short or nonexistent.

2. The *propulsive* phase exists from the time when the center of gravity has passed over the supporting foot to the time when the foot leaves the ground. With the mass center ahead of the driving foot, there is a horizontal-force component vector during the push-off, just as there is a vertical component. The drive is accomplished by knee and hip extension along with plantar flexion of the foot. At this point the body is just barely being supported and there is a gravitational moment turning the body forward and downward around the driving foot. There must be adequate ground reaction against the foot if slipping is to be avoided.

3. The *swing* or *recovery phase* is the period during which the free leg is moving forward to complete the stride. The nature of the leg action during this phase depends upon the speed of the walk or run. In walking, the hip, knee, and ankle must flex just enough to allow the recovery foot to stay clear of the ground. In running, however, the knee flexion increases with speed because the moment of inertia of the free leg tends to resist the muscular efforts of the hip flexors. Therefore, if the knee and hip are flexed considerably, the moment of inertia of the leg is reduced, which simply means that the resistance to angular movement decreases. This same principle governs the extent to which the runner will flex the arms at the elbow. As the stride frequency increases, the arms must move more rapidly to keep up their role in opposition so that good body position can be maintained. Fortunately, such radius shortening of the free leg and the arms occurs naturally during fast walking and running and requires no conscious effort.

The recovery phase is not governed by the law of conservation of angular momentum, because the swinging leg is not an unsupported freely moving segment. The law applies only in the absence of external unbalanced moments. While shortening the radius does tend to increase angular velocity, muscular force is being applied through at least the first half of the recovery action, and the effectiveness of this force is simply enhanced by the reduction of the moment of inertia. It is probable that the last part of the recovery action is ballistic in that the hip flexors develop the leg's forward angular momentum and then relax as that momentum continues until the antagonistic muscles stop and reverse the leg action. But by the time the leg enters the ballistic stage, it has already begun to extend in preparation for ground contact, and so the moment of inertia is then increasing.

Since running is involved in so many sports, some basic understanding of its mechanics is important for coaches and athletes. Yet running is a natural skill, and it is very difficult, perhaps even unwise, to change a runner's style, because of long-standing habit patterns and because of individual physical differences. Corrections may be more feasible in young children who display poor form. One correction that a coach can make is the elimination of any abnormal vertical movement during running. Some vertical motion is unavoidable, because of the oblique-force applications being made, but such movement, if exaggerated, is wasteful of energy and will slow the runner.

In both walking and running, the horizontal velocity is found by multiplying the length of stride by the stride rate. Taller people naturally tend on average to have longer strides. Not every athletic situation calls for the greatest possible speed, but where speed is to be increased, it can be done either by lengthening each stride or by taking more strides per time unit.

Friction and air resistance are the resistive forces to locomotion, and a constant velocity is reached when the propulsive force is just enough to balance the opposing forces. Where applied force exceeds the resistance, acceleration will occur. Obviously a runner who has reached maximum speed has found that the forces opposing motion exactly offset propulsive efforts.

Two sources have compared the movement of the feet in running to the revolving of a wheel. Tricker and Tricker state that "the motion of the foot of a runner follows the motion of a point on the circumference of a wheel surprisingly closely."[2] Similarly, Basmajian has stated, "We are human bicycles. Once rolling on a horizontal surface, we need only small inputs of propulsive force and of balancing mechanisms to maintain forward progress."[3]

Human locomotion over the ground can be of other types, such as leaping, hopping, skipping, sliding, skating, and crawling, but walking and running command the most attention. Steindler provides a comprehensive history of locomotor mechanics and physiological-kinesiological aspects of locomotion.[4] A fascinating collection of photographs was assembled by Muybridge and published in 1887 under the auspices of the University of Pennsylvania.[5] These show not only locomotor action but also a wide variety of other human movements and are worthy of close study by those interested in biomechanics.

[2] R. A. R. Tricker and B. J. K. Tricker, *The Science of Movement* (New York: American Elsevier Publishing Company, Inc., 1967), p. 215.

[3] J. V. Basmajian, "The Human Bicycle," in P. V. Komi, ed., *Biomechanics V-A* (Baltimore: University Park Press, 1976), p. 297.

[4] Arthur Steindler, *Mechanics of Normal and Pathological Locomotion in Man* (Springfield, Ill.: Charles C Thomas, Publisher, 1935).

[5] E. Muybridge, *The Human Figure in Motion* (New York: Dover Publications, Inc., 1955).

Walking and running may be examined through the use of cinematography and videotape, electromyography, force platforms, and computer simulation. They may be viewed from anatomical, physiological, or mechanical perspectives. Finally, attention may be given to running shoes and track surfaces and to their capabilities of receiving and returning energy.[6, 7] Some distance runners have done very well running barefoot, and there is some evidence that, on resilient surfaces, this may be the best way to run.[8]

Throwing

Unlike walking and running, throwing skills do not come naturally to all of us. A fair percentage of adults cannot throw a ball with any efficiency because they either were never taught properly or did not have sufficient opportunities to practice this fundamental skill. The nature of the throwing movement varies widely in sports and includes overhand, underhand, and sidearm patterns as well as throws with one or both hands. The style used depends upon the size and mass of the particular ball, the results desired, and the rules governing the sport. Once the basic throw technique is understood, all variations become clearer. Too much attention has been given in the literature to throws for maximal distance or velocity and to curving pitches. Most throws in competitive and recreational activities probably require accuracy more than they do velocity or distance. Finally, throws can be made to either a stationary or a moving target, as in darts and in football passing, respectively.

For all types of throws, the mechanical considerations include the linear velocity of the projectile at release, the angular velocity of the throwing arm, the height from which the object is released, the angle at which it is released, and the spin that is imparted to the object. These factors determine where, how far, and how fast the projectile will travel.

The overhand throw is perhaps the most useful and common type of throw. A student who learns good form at the beginning need adjust only a few variables to throw to different distances or at different speeds. The ball should be held with the fingers, because the lever length is increased, a higher point of release can be achieved, the wrist and fingers are in a position to contribute the final force, and better control results. The trunk should be rotated toward the ball, and the throwing arm drawn back with the elbow bent. This permits force application over a greater distance, reduces the moment of iner-

[6] G. B. Ariel, "Biomechanics of Athletic Shoe Design," in P. V. Komi, ed., *Biomechanics V-B* (Baltimore: University Park Press, 1976), pp. 361–67.

[7] T. A. McMahon and P. R. Greene, "Fast Running Tracks," *Scientific American*, December 1978, p. 148.

[8] Ariel, "Biomechanics of Athletic Shoe Design," p. 367.

tia of the arm at the start of the throw, and puts the shoulder and anterior chest muscles under stretch, increasing their potential force capabilities. The leg opposite the throwing hand should be ahead of the other leg, in accordance with the principle of opposition, to allow trunk rotation around the hip joint of the leading leg.

The throw begins with a step by the lead leg as the trunk and shoulder rotations initiate the movement (Figure 8-4). The arm is brought around and gradually extends to lengthen the lever and provide for a higher linear velocity of the ball, and finally the flexor muscles of the wrist and fingers complete the summation of forces as the ball is released. Some follow-through is desirable to permit the arm to decelerate smoothly. The ball will leave the hand at a tangent to the arc being described at the instant of release

Figure 8-4 In baseball pitching, the application of a number of mechanical principles is apparent. Here the pitcher is in an excellent position to begin the final delivery motion. *Photograph courtesy of the Ohio State University Athletic Publicity Department.*

(Figure 8–5). If the ball goes higher than the target, the ball may have been released too soon; and a late release might be the cause if the ball falls short of the target. The finger grip on the ball provides the centripetal force to keep it in a circular path.

This description of the basic overhand throw is subject to many variations. For example, in cricket bowling, the pitching arm is kept extended throughout. A baseball catcher needing a quick release for a throw to second base has no time to assume an opposition position or to withdraw the arm very much.

The individual player in any sport has to manipulate the conditions for an effective throw to suit his or her physical capabilities and to achieve

Figure 8–5 As the baseball is released, the pitcher's arm is extended at the elbow and the ball is moving tangentially to the radius. *Photograph courtesy of the Ohio State University Athletic Publicity Department.*

desired results, such as getting rid of the ball in a hurry, getting optimal velocity or distance, or getting accuracy. Thus a baseball infielder must have a command of a variety of throwing styles and speeds to cope with each tactical situation, and a football quarterback must be able to use different arm-movement patterns for short passes, long passes, and pitchouts.

Spin imparted to a ball or other object may be incidental, as in an infielder's throw, or may be deliberate, as in pitching a baseball. In a football pass, a javelin throw, or a Frisbee toss, spin is important for stability in flight; the angular momentum possessed by spinning objects resists changes in orientation.

To achieve higher throwing velocities, the thrower will often take one or two strides before throwing or at least one long stride of the type used by baseball pitchers. This adds to the linear momentum, allows for a longer force application, and flattens the throwing arc, the last reducing the error in release accuracy. According to one source, close to 50 percent of overhand throwing velocity is attributable to the body rotation and preliminary step.[9] Of all the factors involved in distance throws, Kunz states that "in ball and javelin throwing, the velocity that is imparted to the object is the most important factor for maximal distance."[10]

Catching

Whether a catch is made with the bare hand, a gloved hand, or an implement such as a lacrosse stick, a few mechanical principles can be applied. A ball of some mass, velocity, and dimension has a kinetic energy that must be absorbed by the catcher.

The learner has several problems to solve in the very short time that it takes a ball to arrive: judging its speed, aligning the body to receive the ball, keeping the eye on the ball, and properly positioning the hands for the catch. These are especially formidable tasks for a young child, particularly when accompanied by a fear of injury if the ball is missed. Needless to say, use of a beanbag relieves the fear and makes catching considerably simpler to learn.

The key to the painless catching of a fast-moving ball is the reduction of its kinetic energy over as long a distance as possible. This will ensure a

[9] T. Hoshikawa and S. Toyoshima, "Contribution of Body Segments to Ball Velocity during Throwing with Non-Preferred Hand," in P. V. Komi, ed., *Biomechanics V-B* (Baltimore: University Park Press, 1976), pp. 109–17.

[10] J. Kunz, "Effects of Ball Mass and Movement Pattern on Release Velocity in Throwing," in R. C. Nelson and C. A. Morehouse, eds., *Biomechanics IV* (Baltimore: University Park Press, 1974), pp. 163–68.

perfectly inelastic collision; that is, it will have the result that the ball will stay in the hands and the hands will feel the lowest possible average force as they do work to stop the ball. The catcher meets the ball well out in front of the body, and then the elbows flex to ride the ball in toward the chest. Where a glove is worn, the padding absorbs at least some of the ball's kinetic energy, so that the arms need not yield as much. Another factor to consider is the size of the ball being caught. A large ball applies its force over a large area of the hand, which means that there is less pressure and presumably less pain.

In catching a large, heavy object such as a medicine ball, the receiver should stand with his or her line of gravity as near as possible to the ball; this way, the arms can yield and the body's center of gravity can move a fair distance backward as the ball is caught to aid in energy dissipation. If necessary, a step or two backwards can be helpful.

Pushing

Although dissimilar in most respects, the sports of wrestling, shot-putting, football, karate, fencing, boxing, and basketball all involve some pushing motions. Different outcomes are expected in each of these sports, to be sure, but the pushing motion is one in which a linear force is applied away from the body against some object or person.

The shot-putter sequentially applies various forces over as long a distance as possible in order to finally push, or put, the shot straight out from the shoulder. This linear push is, of course, the result of angular movements of the upper arm and forearm; the shot has already received some linear momentum from the athlete's movement across the ring.

The wrestler and the offensive lineman push an opponent, for whatever strategic reason. The longer the pushing force can be applied, the more effective it will be. The opponent can resist the push by spreading his feet in line with the direction of the force, by lowering his center of gravity, or by leaning well forward toward the pusher. However, by leaning forward he is vulnerable to a sudden pull by a clever opponent, because his line of gravity is so close to the forward edge of his support base that even a slight pull can cause him to topple. Because acceleration equals force divided by mass, the acceleration achieved in the pushing motion is indirectly proportional to the mass of the person being pushed. This mass can be effectively increased in a sense if that person digs in with cleated shoes to increase the frictional force against the turf.

In combative sports, a hand is moved forward in a pushing motion until contact is made with an opponent, and so the purpose of the push in this case is not to move weight but to develop kinetic energy and velocity. Some push is accomplished upon contact.

One- and two-hand set shots and passes are essentially pushing actions. In a two-handed pass or shot, the basketball player must be careful to apply equal force with each hand to prevent an unbalanced moment of force that will act to spin the ball around a vertical axis. Where spin is desired in a shot, the push should be made eccentrically by directing the hand force beneath the ball to get backspin for a free throw or to the right or left side to get spin for a bank shot.

HITTING AND STRIKING

In any discussion of hitting and striking skills, a number of principles relating to collisions are of importance. The hitting may be done with the bare or gloved hand, as in boxing, karate, handball and volleyball, or it may be done with a bat, racket, or similar implement. The object struck might be a ball, a shuttlecock, a board, or a body. Each combination of implement used and object struck presents a unique study.

Batting

Batting a ball involves an imperfectly elastic collision in which the ball and bat are briefly in contact and then separate. The ball is temporarily deformed and stores elastic potential energy, which is almost instantly converted to kinetic energy as the ball reforms due to its elasticity. Some energy is lost as heat. If the striking surface is a tennis, racquetball, or squash racket, its strings also deform during the impact.

The length of the bat, racket, or club is one factor in determining the linear velocity of the point where impact occurs. In theory, the longer the implement, the higher the possible linear velocity of the impact point. There is some limit to increasing the lever length, however, because accuracy will suffer as any small error at the hands is magnified several times at the end of the lever. Also, lengthening a lever increases its moment of inertia so that it becomes more difficult to attain a high angular velocity. For this reason, many squash players choke up on the racket to permit quicker wrist action in stroking.

Another consideration in hitting is the stroke pattern used in relation to an incoming ball. Correct judgment of a ball's velocity and path is difficult for beginners, and good hand-eye coordination is needed to time the swing to successfully meet the ball. In a tennis ground stroke, the beginner should be taught to avoid a circular stroke, such as that used in baseball batting, since this allows only one optimal instant for contact with the ball. Chances for good contact are improved by flattening the stroke arc so that the racket moves in a linear path for the several inches where contact is likely.

The law of conservation of momentum states that when two bodies collide, the combined momentums before impact will be unchanged after a perfectly elastic collision. If the kinetic energy that is in fact lost to other forms of energy is ignored, it may be said that any increase in the velocity of a ball after impact is accompanied by a proportionate decrease in the velocity of the bat. The velocity and direction of a ball leaving a bat or racket depend upon (1) the spin, direction, velocity, mass, and elasticity of the ball before impact and (2) the direction, velocity, mass, and elasticity of the bat or racket before impact. In baseball batting, the player may increase the momentum of the bat by swinging it faster, by selecting a heavier bat, or both if possible. The angular velocity of the bat is the more important consideration.

Another aspect of successful ball contact is time. A player in tennis has less time to react at the net than at the base line. A fair number of eye injuries occur as players who are inexperienced or who have slow reflexes attempt to play a net game and cannot always judge a ball's speed or path. For what it is worth, a baseball player who stands to the rear of the batting box has a fraction more time to decide whether or not to swing than one who stands at the front of the box.

If the player holds the bat vertically while waiting for the pitch, the torque is nil at the hands; and if the arms are held close to the body, the total torque acting is quite small. Summation of forces is important in batting. The lead leg generally moves toward the pitcher to overcome inertia and establish a firm base for the swing (Figure 8–6). This is followed by trunk and shoulder rotation and then by an extending of the arms to lengthen the lever and increase the linear velocity of the end of the bat. The wrist action is the final contributor of force at impact. Once a swing motion has begun, there is a very high angular momentum due to the bat's angular velocity and moment of inertia, and checking the swing requires considerable opposing torques.

A final consideration is that of the centripetal force applied by the hands to hold the bat or racket in its circular path. Many batters and racquetball players wear gloves to improve the friction of gripping.

Striking

Ball-hand collisions involve more complexity than do ball-bat types. It is difficult, if not impossible, to determine the coefficient of elasticity between a handball or volleyball and the human hand. Most of the principles that apply to batting also apply to hitting a ball with the hand. The smaller size of the handball as compared with a volleyball means that the pressure is greater and

Figure 8-6 Softball batting requires disciplined control of the sequence of forces being applied. Here the forward stride has been completed and the batter is ready to begin trunk and shoulder rotation. *Photograph courtesy of the Ohio State University Athletic Publicity Department.*

thus there is a need for gloves. Putting spin on the ball requires an eccentric contact of the hand on the ball. The resulting spin then affects the ball's rebound behavior off the wall, the floor, the ceiling, or the opponent's hand in a handball game. Because of the ball's smooth surface and the short time in flight, the curvature of flight path of a spinning handball is of no consequence.

Striking with the hand is done in boxing and some of the martial arts. The collision is between body parts rather than between ball and bat or ball and hand, but many of the same principles apply.

Striking a body part with the hand is not usually classified as an inelastic collision, but it is at the low end of the elastic scale in terms of coefficients. Because every blow is unique, quantitative measures are not feasible. The boxer's jab is more effective if it has a high velocity and if the boxer leans into the punch to provide mass against the reaction to the blow. The left jab is the punch most frequently used by right-handed boxers for two reasons. First, the left glove is typically held farther out than the right when the boxer is on guard so that it has the least distance to travel to reach the target. Second, being a linear movement, the jab takes less time to arrive than does the more powerful hook, which follows a circular path.

As in catching, the receiver of a blow can yield the part being struck (roll with the punch) and thereby increase the distance over which the kinetic energy of the punch is dissipated. The padded glove itself absorbs some of this energy and also spreads the force over a large area of the body part being hit. Beginning boxers should use 16-oz gloves for this reason. Boxing headgear can be worn, but it tends to give partners a false sense of security against hard blows.

Although the bare-handed breaking of boards and concrete blocks is not a primary emphasis in karate, it is one of the more dramatic and publicized aspects of the sport (Figure 8–7). One wonders how the human hand, considering its fragile structure, can be used almost as a hammer without sustaining injury. Intense mental concentration is certainly a factor, but mental skills do not negate mechanical laws. For every action there is an equal and opposite reaction.

This type of breaking can be classified as an inelastic collision. The hand velocity and thus the momentum are very high at impact, and the force is applied to a very small surface area of the block or board. According to Feld, human bone can resist 40 times more stress than concrete, and the impact force needed to break concrete is about 3000 N.[11] The stress of impact is transmitted from body part to body part beginning with the skin at the point of contact and moving through muscle, tendon, and tissue all the way up the arm.

Kicking

Kicking can be grouped with hitting and striking and perhaps even with throwing because of a number of similarities in the ruling mechanics. All these skills involve the development of an angular momentum obtained as the product of a limb's angular velocity and its moment of inertia. The linear velocity of the kicking foot is the product of the effective radius and the leg's

[11] M. S. Feld, R. E. McNair, and S. R. Wilk, "The Physics of Karate," *Scientific American,* April 1979, pp. 150–58.

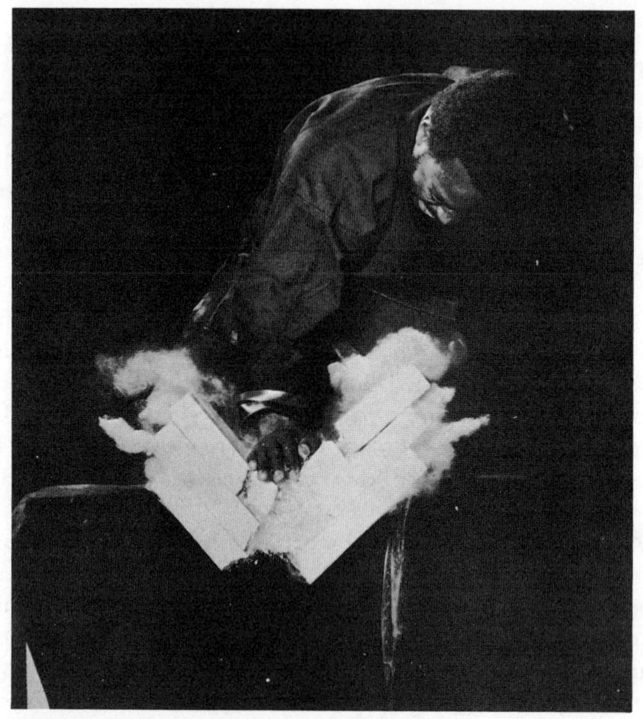

Figure 8-7 The breaking of concrete blocks with the hand requires application of force approaching 3000 N. *Photograph courtesy of Dr. Ronald E. McNair.*

angular velocity at impact. As with batting, most kicks entail an elastic collision with a ball.

Where in batting a final acceleration is given to the implement by the wrists, in some kicking skills the lower leg accelerates just before contact. This acceleration is due in part to the action of the quadriceps group, but also in part to the sudden slowing of the upper leg's angular momentum, which is then transferred to the lower leg.

As in throwing, a kicker often has a high degree of control over the spin that is imparted to the ball. This is especially true when the ball is stationary before the kick, as it is in field goal kicking and in soccer penalty kicking. An off-center kick will give the ball spin, with backspin generally desired for distance kicks.

It would be very difficult to calculate a coefficient of elasticity between the kicking foot and a ball, but the inflation of a ball is critical, and of course the pressure must be within the rules specifications. Even a properly inflated ball will deform upon foot contact, and its elastic reformation is one factor in flight distance. Direction and distance are also governed by the force of the kick, the point of application of the foot on the ball, and the direction of the applied force vector. Someone doing a kicking analysis would utilize those principles that directly relate to the type of kick. There are similarities and differences between punting and place-kicking and between kicking a stationary soccer ball for distance and passing an already rolling soccer ball.

JUMPING

There are a number of different types of jumps, and each may be used in one or more ways. A vertical jump can be made with or without an approach run. A long jump may be made from a stationary position or after a run. The takeoff might be from one or both feet. It might be used to clear an obstacle or to jump over a rope. Whether it is further distinguished by being called a hurdle, a hop, or a leap, a jump is any sort of spring into the air as a result of leg-muscle action. The principal jumping muscles are the hamstring group and the gluteus maximus, which extend the hip; the quadriceps femoris, which extends the leg at the knee; and the gastrocnemius and soleus, which plantarflex the foot at the ankle.

Jumping, however low, short, high, or far, is a study in projectile motion. Unless the jump is vertical, there are two velocity components to consider, horizontal and vertical. Time in the air is a function of the vertical velocity component ($T = 2v_y/g$). Height reached is related to vertical velocity and time ($s = \frac{1}{2}gt^2$ or $v_y^2/2g$). Distance is the product of horizontal velocity and time in the air ($d = v_xT$).

A jumper's ability or style may be limited by rules of the sport or by the laws of physics. Of the several laws that apply, Newton's second law is common to all jumps, and it states that acceleration equals force divided by mass. Thus, where two athletes have equal force-producing potential in their legs, the lighter person will experience the higher acceleration, and other things remaining equal, this higher acceleration will mean a higher takeoff velocity and therefore a higher or longer jump. Weight training for the legs can increase the available force, and proper technique should result in the efficient summating of several forces.

Vertical Jumping

The ability to jump vertically is necessary in a number of sports. A one-footed takeoff is used in a basketball layup shot and by an end in football when catching a high pass. A two-footed takeoff is used in basketball for the center-jump and the rebound; in volleyball for blocking and spiking; in dance; and in rope jumping. While the effective use of the arms can add a few inches to the height of a jump, priority must be given to positioning the arms and hands for a particular action such as blocking, spiking, catching, or rebounding a ball (Figure 8–8). A vigorous upward swing of the arms followed by their rapid deceleration will allow for transfer of momentum to the whole body and will reduce the amount of inertial resistance to the lifting force of the legs.

According to Margaria, the legs can exert greater force in a vertical jump if there is a preceding bounce to put the muscles under stretch and "load" them with elastic energy.[12] This is a common technique in basketball and volleyball, where maximum height is often necessary.

Figure 8–8 Volleyball blocking and spiking require exceptional vertical jumping ability. *Photograph courtesy of the Ohio State University Athletic Publicity Department.*

[12] Rodolfo Margaria, *Biomechanics and Energetics of Muscular Exercise* (Oxford: Clarendon Press, 1976), p. 119.

Rope jumping, by contrast, requires the lowest possible jump, just enough to clear the rope as it passes under the feet. It can be done with both feet, one foot, or alternating feet and is generally done on the balls of the feet. Hence, the plantar flexors of the ankle are strongly active concentrically and eccentrically. Endurance is the primary objective, both muscular and cardiorespiratory. The hands and wrist get some exercise in the process, but their movement should be minimal. The weight and firmness of the rope ought not be neglected, because at slow turning speeds a light rope tends to be limp at the top of the swing, making centrifugal force low.

The male performer in ballet and modern dance is often required to make high vertical jumps from a stationary stance. Where such jumps are accompanied by one or more turns in the air, he must apply both a linear vertical force and an eccentric force around his longitudinal axis. The turn must be initiated while he is still in contact with the floor.

Standing Broad Jump

The standing broad jump is widely used in physical fitness testing as a measure of leg power in children. As in all jumping, the key is Newton's second law, but since there is no pressure of time to complete this type of jump, more effort can be devoted to efficiently summing the body forces. The usual preparatory action involves rocking back and forth once or twice, shifting the weight from the toes to the heels as the arms move from flexion to hyperextension. As preparation ends, the line of gravity is near the heels, the hips and knees are flexed, the ankles are dorsiflexed, and the arms are hyperextended.

The propulsive phase is a composite movement of the center of gravity forward beyond the toes and the beginning of the shoulder flexion as the arms swing in a large arc in the sagittal plane. The angular momentum developed by the arms abruptly ends just as the leg drive begins. When this happens the momentum of the arms is transferred to the rest of the body so that the legs have less weight to push. With the center of gravity well ahead of the toes, the vector of the leg force passes through the center of gravity to provide a projection angle that will permit the vertical component of velocity to lift the body into the air so that the horizontal component is given time to act effectively.

As soon as the body is airborne, the legs quickly position themselves for landing, at which time they absorb the kinetic energy of the jump through flexion of the major joints of the legs and hips. It is important to provide a firm landing surface with good traction; otherwise the jumper's extended legs would cause the feet to slip forward on account of insufficient frictional reaction.

Running Long Jump

The running long jump, as contrasted with a standing jump, relies more than anything else on the athlete's linear velocity in the approach. Most long-jumpers are also sprinters, but speed alone will not result in good jumping distance. The athlete must apply an impulse to the takeoff board to change direction from horizontal to some angle of projection. Because of the retarding effect of the takeoff foot on the board, some horizontal velocity is lost while a vertical component is acquired.

The faster the athlete's approach run, the less the time on the board to change direction. Bedi and Cooper[13] found that the foot-to-board contact time was 0.11–0.14 s when the mean approach velocity was 26.6 ft/s, with the better jumpers on the board 0.11–0.12 s.

It is not possible following a fast run to project the body at an ideal 45°, but the larger the angle of takeoff, other things being equal, the farther will be the jump. A variety of angles have been reported in studies, and the range appears to be between 17° and 26°.

In the final step of the approach, the center of gravity is slightly lowered. The contacting foot is ahead of the center of gravity, and as it strikes the board there is a retarding reaction that has the result that a slight forward angular momentum is imposed on the athlete.

Once airborne, the jumper typically takes a stride in the air, known as the hitchkick, in order to remain erect and get the legs into a good landing position. These efforts do not affect the path of the jumper's mass center, as only an outside force has any effect on a projectile's parabolic path.

In a perfect landing, the jumper's feet touch side by side and the jumper falls forward. Measurement is from the point of body contact nearest the takeoff board. Flexion at the hips and knees is necessary to absorb the energy of the jump without joint injury.

High Jump

In high-jumping the athlete's approach is not nearly so fast as that of the long-jumper, for the obvious reason that a radical change in direction must be made at takeoff, which would not be possible at high speeds. The long-jumper seeks maximum horizontal distance where the high-jumper needs only enough horizontal velocity to carry the body past the bar after it has been cleared.

There are two styles in current use, the straddle and the more

[13] J. F. Bedi and J. M. Cooper, "Take-off in the Long Jump: Angular Momentum Considerations," *Journal of Biomechanics,* vol. 10, 1977, pp. 541–48.

popular Fosbury flop. While the styles are quite different, there are a number of common elements. Maximum efficiency is demonstrated when the jumper's center of gravity is raised no higher than necessary to clear the bar. This is enhanced by the lifting of the arms and the free leg before ground contact is broken so that the jumper's center of gravity is as high as it can be before takeoff. In this regard the taller jumper has some advantage, although there are infrequent exceptions to this rule, of which the most recent is Franklin Jacobs, who at 5 ft 8 in. cleared 7 ft 7¼ in. in 1979. In both styles, a long final approach step allows the longest possible ground contact time during which to apply a vertical force, that is, to convert the horizontal velocity to a vertical velocity. This long step also allows the free leg time to swing and develop angular momentum along with the upswinging arms. As the arms and free leg reach their upward limits, they rapidly decelerate at the instant the supporting leg applies its force eccentrically to provide the body rotation needed to permit effective clearing of the bar (Figure 8-9).

Figure 8-9 An example of the Fosbury style of high-jumping. *Photograph courtesy of the Ohio State University Athletic Publicity Department.*

In the straddle style, now seldom used by women jumpers, the athlete rolls over the bar because of the angular momentum received at takeoff. The flopper uses a faster, semicircular approach and takes off facing away from the bar, using the angular momentum obtained by the free leg's motion up and diagonally across the body.

Once in the air, the jumper is subject to the law of conservation of angular momentum, and both the path of the body's center of gravity and its axis of angular momentum have been established. But to clear the bar, the jumper can use the law advantageously; movement of a segment in one direction around the axis will result in an opposite movement of some other segment. Every motion in the air must have a purpose if bar contact is to be avoided.

Although the manner of landing does not affect the results in high-jumping as it does in long-jumping, safe landings are of some concern. Today's seven-foot jumps might not have been possible with the ground-level sand or sawdust pits used some years ago. It is unlikely that anyone would risk a broken neck using the Fosbury jumping style, which requires a landing on the back of the shoulders and neck, had the new foam rubber elevated pits not been introduced. The modern high-jump pit reduces the velocity with which the jumper hits and provides a long distance over which to absorb the energy.

WEIGHT LIFTING

The lifting of weights, whether for competitive or training purposes, requires a knowledge of anatomy and mechanics if maximum strength or muscular endurance benefits are to be derived with the least chance of injury to joints or tissue. The techniques of competitive weight lifting are well-defined both by rule and by practice. Noncompetitive lifting is called *weight training* and is used by many male and female athletes as a means of strength development for sports. Weight training should be specific to the sport and even to the individual, because no single program is suitable to all sports or to all athletes.

Lifting of free weights exercises a great many more muscles than most athletes realize. From a standing position, a press type of lift, done with the intent to strengthen arm and shoulder muscles, requires muscle contractions at many joints, which need to be stabilized for the lift to be successful. While the arms are pressing the barbell upward, the entire body experiences the added weight, and accordingly, there must be a series of joint-stabilizing contractions in reaction to the weight borne above.

Whole-body involvement is reduced by lifts that are done on the various universal gyms, strength machines, and isokinetic apparatus that allow for concentration of effort on a particular muscle group (Figure 8–10).

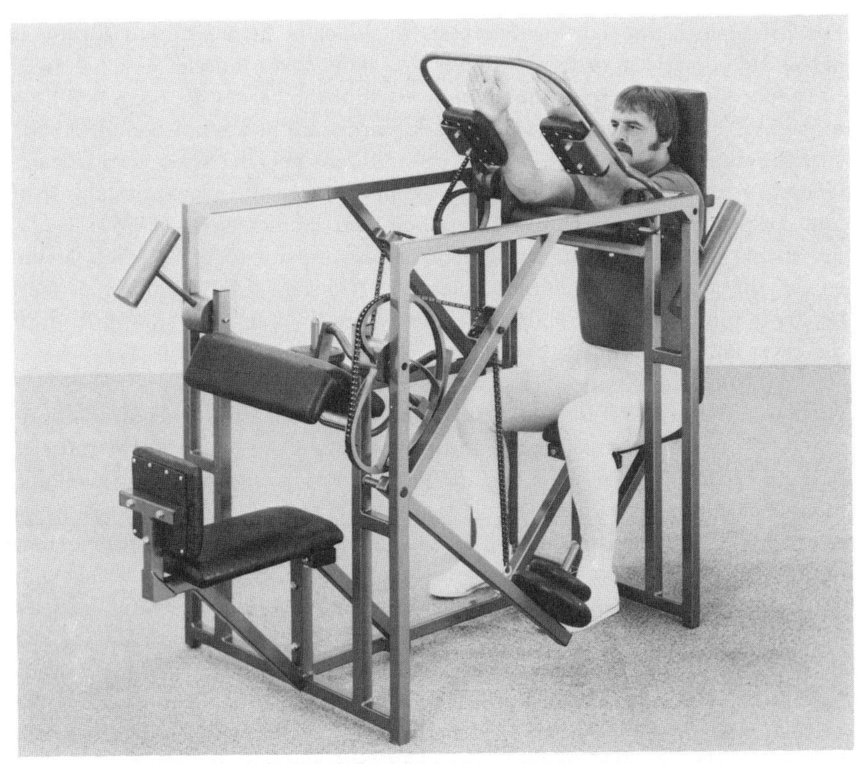

Figure 8-10 An example of a machine designed to develop the elbow flexor muscle group over a full range of motion. *Photograph courtesy of Nautilus Sports/Medicine Industries.*

The nature of some of this equipment probably reduces the likelihood of muscle injury during exercise.

It is, or should be, common knowledge that back injuries in heavy lifting can be minimized by initiating the lift with the strong muscles of the legs while maintaining a fairly erect spine. As the bar's inertia is thus overcome by the straightening of the legs and before the bar has stopped moving upward, the shoulder and arm muscles complete the lift. Nachemson found that there was 250 kg of pressure at the third lumbar disc when a 20-kg weight was lifted with bent knees and a vertical back, as compared with 380 kg of pressure in lifting the same mass with the knees straight and the back flexed.[14]

During any lift, the weight should be kept as close as possible to the body so as to reduce the external moment opposing the lift. An exception to

[14] A. L. Nachemson, "Low Back Pain: Its Etiology and Treatment," *Clinical Medicine,* January 1971, pp. 18–24.

this principle is the curling exercise, in which a high external torque is deliberately sought in order to provide the elbow flexor muscles with a desired resistance. In this exercise the actual weight of the barbell is magnified by a moment arm that is longest when the forearms are at the horizontal. As the curl begins with the weight in front of the thighs and the arms vertically extended, there is almost no turning moment, because the gravitational line of force passes through or close to the elbow and shoulder axes (Figure 8–11). From this initial position, through a curling arc up to 90° flexion at the elbow, the gravitational moment increases to its maximum value and then begins to diminish to the end of the curl, when the bar is directly in front of the chest and the torque is again near zero.

In curling, some mechanical and anatomical advantages and disadvantages are traded off in the course of the movement. In the beginning position, there is no external torque to deal with, but the elbow flexor muscles have an extremely poor angle of pull to start rotation, as most of the available force is expended as a compression component. Hence the muscles are in an inferior position when the stationary inertia of the barbell makes movement difficult. This disadvantage is somewhat offset anatomically by the fact that the muscles being stretched can exert more force. Now, as the muscles shorten and the curl progresses, the external moment increases, while the shorter muscles can exert gradually less force. Fortunately, a larger proportion of the muscle force is available for rotation because of the more favorable angle of attachment of tendon on bone, which, together with the inertia of the already moving weight, makes it easier to keep the barbell in motion.

The muscles that contract concentrically to raise the weight through

Figure 8–11 In curling a barbell, the external moment is greatest when the forearms are horizontal.

positive work are the same as those that eccentrically contract to lower the weight which does negative work on the muscles. Because strength gains therefore occur in both directions of the curling movement, the weight should be lowered under muscular control and should not be allowed to simply drop back to the starting position.

Curling is a good exercise for developing the elbow flexors as well as the flexor muscles of the hand and wrist, many of which are located in the anterior forearm. An individual should do the curl in a uniform way every time, because the purpose of the exercise is to develop particular muscles and not just to move a weight. As fatigue sets in, there is often a tendency to cheat by thrusting the hips forward to help initiate the curl or by leaning backward during the curl to reduce the external moment arm. While these tactics may mentally satisfy the lifter who has in mind the completion of a certain number of repetitions of a given weight, they do not serve to concentrate the effort on the strengthening of the muscles in a systematic way.

Another mechanical consideration involves the construction of the barbell. Most commercially made barbell sets allow the plates to rotate rather freely around the bar. This is desirable because, when a lift is made in any manner that involves an angular displacement, the plates have a moment of inertia resisting any change in their state of angular motion. Therefore the bar, being gripped by the hands, will undergo some angular displacement, as will the plates, but the plates will not rotate along with the bar. Were the plates to be welded to the bar, they would rotate and develop an angular momentum and an angular kinetic energy, which would require the hands to squeeze harder at the completion of a lift to stop the angular momentum. While this presents no great problem with light weights, a heavy weight would have a definite tendency to roll right out of the lifter's grasp.

With the heavier weights, proper hand spacing is very important if the torques acting at either end of the bar are to be balanced. In some lift styles such as the snatch, the hand spread is very wide to permit the arms to straighten without having to lift the weight as high as they would with the hands closer together. This also reduces the external moments acting at the ends.

GYMNASTICS

Because of the speed and complexity of modern competitive gymnastic movements, no attempt will be made here at anything approaching a complete analysis. This sport is ideal for the application of a great many biomechanical principles, and what follows is an elementary discussion of some of the more simple aspects of gymnastics.

Rotatory motion is nearly always involved in gymnastic skills, and so the coach and performer should be familiar with the laws of angular motion. First of all, to initiate rotation or twist, an eccentric force must be applied before the performer has left the supporting surface or apparatus. In tumbling, this is accomplished by positioning the body so that the leg-force vector does not pass through the mass center of the tumbler (Figure 8–12). The resulting moment of force will produce an angular momentum that, once established in an airborne performer, will not change in magnitude or direction. This is in accordance with the law of conservation of angular momentum ($A = I\omega$). Although the angular momentum of an airborne, rotating gymnast cannot be changed, his or her angular velocity can be increased or decreased by altering the moment of inertia. The athlete does this by manipulating body parts around the axis. From a layout position, for example, the gymnast's angular velocity will increase if a tuck position is taken, because the moment of inertia of the body will have been reduced to as little as a fourth of its layout value. Similarly, if a somersaulting performer in tuck position opens up into a layout, the increased radius will result in a dramatic reduction in the angular velocity, but the total angular momentum will have been conserved.

Having established an angular momentum around some axis, a trained gymnast can change the axis and move, for example, from a somer-

Figure 8–12 To perform a somersault, the performer must initiate rotation while still in contact with the supporting surface or apparatus. If the athlete leans slightly forward at the time the legs are extending, the force vector passes behind the mass center and creates a moment of force that will rotate the performer.

sault into a twist. This should not be construed as the initiation of a new rotation, because it is simply a matter of utilizing an existing angular momentum in a different way, a trading of rotation around one axis for rotation around another.

Performers on the horizontal bar, uneven parallel bars, and rings must contend with centripetal forces, and a beginner soon learns that a strong grip is needed to keep from flying off during various high-speed maneuvers. The hands provide the centripetal force necessary to cope with the body's mass, velocity, and effective radius as the gymnast goes through a circular motion whose axis is at the hands. The need for centripetal force is greatest at the bottom of any swinging movement and is least at the top where the body weight acts downward through the grip. It is for that reason that the horizontal-bar performer in Figure 8–13 waits until he is at the top of a giant swing before he makes any grip change.

In the various bar and ring events, there is a continuous interplay between kinetic energy and potential energy. Potential energy is gained on the rise and is reduced as the body falls. Because gravity will naturally slow the body during the upward phase of a swing, the gymnast subtly shortens the swing radius and thereby increases the body's angular velocity. Speed is gained on the downswing when the body is fully extended to provide the largest gravitational moment. A woman performing on the uneven parallel bars deliberately collides with the lower bar and circles it as part of her routine. Having had one axis in the form of the bar she was gripping, her hips strike the lower bar, which then takes on the role of the axis. As this hip-bar contact is made, the momentum of the whole body is transferred to the legs, which therefore experience angular acceleration as the performer assumes something of a pike position. If she lets go of the upper bar, her whole body rotates around the lower bar, but if she retains her grip, the leg swing will stop and be reversed as in a delayed elastic collision. At its higher levels, this event is one of the most intricate in all sports.

DIVING

Springboard and platform diving share many movements with gymnastics, and therefore many of the same principles can be applied. The characteristics of the diving board and the actions of the diver during the approach, hurdle, and takeoff determine what the diver will be able to do in the air. When a man completes his hurdle on the end of the board, he has done work to depress the board and give it an elastic potential energy. This energy is then directed upward to provide both a lifting impulse and a reaction force to the diver's own leg drive. If the diver is a beginner, he may sometimes leave the board before it has had time to do the work in lifting him.

Figure 8-13 At the top of a giant swing, the centripetal force needed is less than at any other point in the swing, and it is here that the gymnast makes any grip changes. *Photograph courtesy of the Ohio State University Athletic Publicity Department.*

The upward acceleration the diver will experience is a function of the depth of board depression and of the diver's mass, leg-extension strength, and skill at timing his movements to take maximum advantage of the board's action. As with any projectile, the diver cannot change the path of his center of gravity once he has left the board.

All dives require some amount of angular momentum at takeoff. An eccentric force must be applied by the board to the diver as it lifts him, which means that the board's force vector must be directed behind the diver's center of mass to initiate front somersaults or front dives and to the front of the mass center to rotate the diver for a back dive. This is accomplished either by the diver's leaning as the board lifts him or by his applying a forward or rearward force component with his feet against the board.

The amount of lean from the vertical may vary from 14° for a jack-knife type of dive to 27° for a two-and-one-half somersault. One of the problems a beginner must solve is the proper amount of lean for successfully doing a simple front dive. Fear of hitting the board on the way down causes many novices to lean forward excessively so as to get away from the board. They usually succeed in avoiding board contact but at the cost of gaining so much angular momentum that they complete a three-quarter somersault and land on their backs. On the other hand, not leaning enough might result in a prone landing instead of the desired head-first entry.

As with gymnasts and trampolinists, the airborne diver can lengthen or shorten the body's radius to slow or increase the rate of spin. The law of conservation of angular momentum indicates that a freely revolving body maintains its angular momentum in the absence of external moments of force. Thus a diver who changes from the layout position to the tuck position will reduce the body's moment of inertia and achieve an equivalent increase in angular velocity. A change from layout to pike reduces the moment of inertia by about one half, while a change to tuck reduces the moment of inertia to approximately one fourth of what it is in the layout position. The highest angular velocities in diving are achieved in twists around the longitudinal axis when the diver brings the arms in close to the body.

The somersaulting diver who opens up just before entry is actually still rotating, but in contrast to the rapid spinning which has preceded, the now extended position gives the illusion of being an entirely linear motion. Of course, the timing of the opening is such that entry into the water immediately follows, and there is no time for the continued rotation, however slow, to be apparent.

While it is *possible* for an experienced diver to initiate some rotation in the air, correct and efficient diving requires that rotations be started while the diver is still in contact with the board. After all, a particular dive is planned and not improvised in the air.

BOWLING

Bowling is not a mechanically complex sport, but a number of principles can be applied. There is interaction between the bowler and the ball, between the ball and the alley, and between the ball and the pins.

In the preparatory stance, the ball is supported by both hands and is held close to the chest so that there is very little torque. As the approach begins, the ball is pushed forward well out in front of the advancing body, and this creates a long moment arm and a high moment acting on the ball. The ball has potential energy, and the gravitational moment initiates the backswing.

The bowler actually walks forward past the rearward-moving ball; there is no muscular force being applied by the arm. As the ball and arm reach the end of the pendular motion, the ball again has potential energy, and its upward motion is stopped by the force of gravity and some effort by the antagonistic muscles of the shoulder.

A high backswing is probably desirable and is characteristic of better bowlers who are able to deliver the ball at high velocities. However, a study by Mursae indicates that the ball velocity was not significant in scores achieved by subject bowlers.[15]

The delivery phase takes place during the final stride while the body is lowered enough to flatten the arc of the ball and to permit release close to the floor. As with any object held in a circular path, the ball will travel at a tangent to the arc being described at the instant of release. Therefore, the bowler's arm should be just about vertical at delivery. Holding the ball until the arm is beyond the perpendicular will cause a lofting of the ball, a typical error of the beginner. The centripetal force, which has been applied throughout by the fingers gripping the ball, is removed, and the delivery is often made with an eccentric force applied to the ball to cause it to spin in the transverse plane around a longitudinal axis. Whether with a flat release or some spin, the ball will skid a few feet and then begin to roll as friction acts between the ball and the floor to provide an eccentric force. Such rolling occurs sooner for low-velocity deliveries typical of novices.

Everything else being equal, a heavy ball should knock down more pins than a light ball. The law of conservation of linear momentum applies in the nearly perfectly elastic collision of balls and pins and between the pins themselves. Whatever momentum is gained by the toppled pins is lost by the ball, so that the total momentum is nearly the same as it was before collision.

SWIMMING

Scientists have studied the movement of fish and mammals through the water and of ships on the water, but the intensive study of human swimming abilities is still in its infancy. The areas of interest are the starts, the turns, the reduction of resistance, the maximization of propulsive force (stroke mechanics), and the design of pools, starting blocks, and lane lines.

Karpovich conducted one of the earliest research projects on swim-

[15] Y. Mursae and others, "Biomechanics of Bowling," in R. C. Nelson and C. A. Morehouse, eds., *Biomechanics IV* (Baltimore: University Park Press, 1974), pp. 291–97.

ming and water resistance.[16] The resistance aspect of swimming continues to receive attention, but the reduction of resistance probably does not hold as much hope in improving swimming times as does increasing propulsive force. As with runners, swimmers can do very little to reduce body resistance beyond wearing smooth, form-fitting suits. They can increase muscular strength and improve muscular endurance through training. Very positive contributions to lowering competitive times have been made by pool and lane line designers in minimizing the waves caused by swimmers.

The value of the grab start has been verified by, among others, Bowers and Cavanagh, who found that the time gain in this style came while the swimmer was still on the block and was presumably because the swimmer could hold the marks steadily while tensing the leg extensors.[17] A variety of other studies regarding starting can be found in the literature focusing on such aspects as takeoff angles and velocities, heights and angles of blocks, and comparisons of starting arm techniques.

One of the earlier reports on kicking forces was by Cureton,[18] and the propulsive capabilities of variations of flutter kicks, dolphin kicks, whip kicks, and frog kicks have long been of interest to coaches and swimmers. Holmér indicates that there is evidence that the crawl flutter kick is not very efficient and should be reduced in the longer swims. He states that the swimmer should kick only enough to balance the body and keep the legs near the surface.[19]

One of the more intriguing concepts of recent years has been the hypothesis that the swimmer's hand behaves as an airfoil in obtaining lift.[20, 21] The contention is that the hand, acting much the same way as does a ship's propeller, must be pitched to get lift and must move in a pattern that continually seeks out still water. This tends to explain why underwater movies of swimmers show the hands to be moving in S-shaped patterns rather than in straight pulls.

It is difficult to study swimmers without putting them in unnatural situations that immediately reduce the likelihood of obtaining valid results.

[16] P. V. Karpovich, "Water Resistance in Swimming," *Research Quarterly,* vol. 4, October 1933, pp. 21–28.

[17] J. E. Bowers and P. R. Cavanagh, "A Biomechanical Comparison of the Grab Start and Conventional Sprint Starts in Competitive Swimming," in L. Lewille and J. P. Clarys, eds., *Swimming II* (Baltimore: University Park Press, 1975).

[18] T. K. Cureton, "Mechanics and Kinesiology of Swimming," *Research Quarterly,* vol. 1, no. 4, December 1930, pp. 87–121.

[19] I. Holmér, "Efficiency of Breaststroke and Freestyle Swimming," in Lewille and Clarys, eds., *Swimming II,* p. 134.

[20] G. W. Rackham, "An Analysis of Arm Propulsion in Swimming," in Lewille and Clarys, eds., *Swimming II,* p. 174.

[21] R. M. Brown and J. E. Councilman, "The Role of Lift in Propelling the Swimmer," in J. M. Cooper, ed., *Selected Topics on Biomechanics* (Chicago: The Athletic Institute, 1970), pp. 179–88.

Swimming subjects have been placed in a variety of test tanks, have had harnesses placed on them, have been filmed extensively from the side, above, and below, and have had telemetering instruments attached to them. With all of that, as necessary and promising as it is, records continue to be broken by swimmers who are simply training harder than ever.

FENCING

Fencing is a very old sport that evolved directly from real swordplay. Therefore, most of the skills employed have been developed to a high level of efficiency and effectiveness after centuries of trial and error. In this combat sport, the most important single movement is the lunge, which is designed to deliver a touch on an opponent who is otherwise out of reach (see Figure 8–14).

Variations of the lunge are found in areas such as dance and even baseball, where a first baseplayer reaching for a wide throw is, in fact, executing the lunge movement. A well-made fencing lunge is fast and covers the

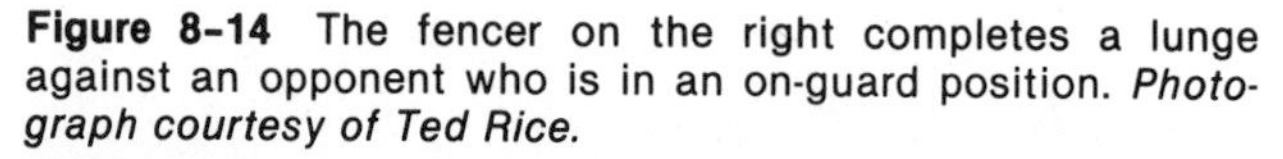

Figure 8–14 The fencer on the right completes a lunge against an opponent who is in an on-guard position. *Photograph courtesy of Ted Rice.*

necessary distance to the target, but it must also terminate with the fencer in a stable position so that a recovery is possible should the hit not be made.

From a flexed-knee on-guard stance, the lunge begins with a simultaneous stepping forward with the lead foot and extending of the rear leg at the hip and knee to provide the force for the lunge. This violent rear leg extension requires good traction between the rear foot and the floor (Newton's third law), because that is the only surface contact point against which force can be developed. Without good traction and correct foot placement, a fencer may easily slip during this critical interval and the lunge will fail. The same problem exists as the lunge is completed; the lead foot must not slide upon landing, because the fencer may fall or pull a muscle.

The kinetic energy produced in the lunge must be absorbed over a rather short distance when the lunge ends. The braking force is supplied by friction between the front foot and the floor and by the eccentric contraction of the extensor muscles of the hip and knee of the lead leg.

There is a continuous diminishing of the angle formed by the force vector and the floor surface, and the horizontal component of the force increases as the angle decreases. Similarly, the relationship between the force vector and the fencer's mass center changes throughout the lunge, with the center of mass lowering as the lead foot moves farther forward.

As the lunge begins, the forward motion of the lead foot causes the body to fall forward around the axis formed by the rear foot. At the same moment, the rear-leg drive provides the impulse to move the body mass toward the target. A moderately deep on-guard stance places the mass center in a position which is nearly in line with the rear leg's force vector and which is low enough that the horizontal force component is as large as practicable.

Many textbooks state that the rear foot must remain flat and in place throughout the lunge. While this may have some value in training the beginner, it is seldom practiced in competition. At some point in the late stages of the lunge, the extensor muscles have already made their maximum force contribution and the action-reaction effects have already been realized. Therefore, it cannot matter whether the rear foot remains flat, rolls, or slides, so long as the fencer can remain balanced at the end of the lunge. Allowing the foot to slide or to roll may permit a slightly longer lunge.

SUGGESTED READINGS

ALLEY, L. E., "An Analysis of Water Resistance and Propulsion in Swimming the Crawl Stroke," *Research Quarterly,* vol. 23, October 1952, p. 253.

BATTERMAN, C., *The Techniques of Springboard Diving* (Cambridge, Mass.: MIT Press, 1968).

BORMS, J., R. MOERS, AND M. HEBBELINCK, "Biomechanical Study of Forward and Backward Giant Swings," in P. V. Komi, ed., *Biomechanics V-B* (Baltimore: University Park Press, 1976), pp. 309–13.

BOWERS, J. E., and P. R. CAVANAGH, "A Biomechanical Comparison of the Grab and Conventional Sprint Starts in Competitive Swimming," in J. P. Clarys and L. Lewille, eds., *Swimming II* (Baltimore: University Park Press, 1975), pp. 225–32.

BRODY, H., "Physics of the Tennis Racket," *American Journal of Physics,* vol. 47, June 1979, p. 482.

BUCHER, W., "The Influence of the Leg Kick and the Arm Stroke on the Total Speed during the Crawl Stroke," in J. P. Clarys and L. Lewille, eds., *Swimming II* (Baltimore: University Park Press, 1975), pp. 180–87.

BUNN, JOHN, *Scientific Principles of Coaching* (Englewood Cliffs, N.J.: Prentice-Hall, Inc., 1972).

CLARYS, J. P., and J. JISKOOT, "Total Resistance of Selected Body Positions in the Front Crawl," in J. P. Clarys and L. Lewille, eds., *Swimming II,* (Baltimore: University Park Press, 1975), pp. 110–17.

DYSON, G., *The Mechanics of Athletics* (London: University of London Press Ltd., 1967).

ECKER, T., "The Fosbury Flop," *Athletic Journal,* vol. 49, April 1969, p. 67.

FOLEY, C. D., A. D. QUANBURY, and T. STEINKE, "Kinematics of Normal Child Locomotion," *Journal of Biomechanics,* vol. 12, 1979, pp. 1–6.

FROHLICH, C., "Do Springboard Divers Violate Angular Momentum Conservation?" *American Journal of Physics,* vol. 47, July 1979, pp. 583–92.

HAY, J. G., *The Biomechanics of Sports Techniques,* 2nd ed. (Englewood Cliffs, N.J.: Prentice-Hall, Inc., 1978).

——, "The Hay Technique: Ultimate in High Jump Styles?" *Athletic Journal,* vol. 53, March 1973, p. 46.

JAMES, S. L., and C. E. BRUBAKER, "Running Mechanics," *Journal of the American Medical Association,* vol. 221, August 1972, pp. 1014–16.

KARPOVICH, P. V., "Analysis of the Propelling Force in the Crawl Stroke," *Research Quarterly,* vol. 6 supplement, May 1935, p. 49.

KIRKPATRICK, P., "Batting the Ball," *American Journal of Physics,* vol. 31, 1963, p. 606.

LIPETZ, S., and B. GUTIN, "An Electromyographic Study of Four Abdominal Exercises," *Medicine and Science in Sport,* vol. 2, Spring 1970, pp. 35–38.

MILLER, D. I., "A Comparative Analysis of the Take-off Employed in Springboard Dives from the Forward and Reverse Groups," in R. C. Nelson and C. A. Morehouse, eds., *Biomechanics IV* (Baltimore: University Park Press, 1974), pp. 223–28.

OFFENBACHER, E. L., "Physics and the Vertical Jump," *American Journal of Physics,* vol. 38, 1970, p. 614.

PLAGENHOEF, S., *Patterns of Human Motion* (Englewood Cliffs, N.J.: Prentice-Hall, Inc., 1971).

RAMEY, M. R., "Significance of Angular Momentum in Long Jumping," *Research Quarterly,* vol. 44, December 1973, pp. 488–97.

ROBERTS, E. M., and A. METCALF, "Mechanical Analysis of Kicking," in J. Wartenweiler, E. Jokl, and M. Hebbelinck, eds., *Biomechanics: Technique of Drawings of Movement and Movement Analysis* (Basel, Switzerland: S. Karger AG, 1968).

ROOZBAZAR, A., "Biomechanics of Lifting," in R. C. Nelson and C. A. Morehouse, eds., *Biomechanics IV* (Baltimore: University Park Press, 1974), pp. 37–43.

SCHLEIHAUF, R. E., JR., "A Biomechanical Analysis of Freestyle," *Swimming Technique,* vol. 11, Fall 1974, pp. 89–96.

SHANEBROOK, J. R., and R. D. JASZCZAK, "Aerodynamic Drag Analysis of Runners," *Medicine and Science in Sports,* vol. 8, Spring 1976, pp. 43–46.

SIMONIAN, C., *Basic Foil Fencing* (Dubuque, Ia.: Kendall/Hunt Publishing Company, 1976).

STEINDLER, A., *Mechanics of Normal and Pathological Locomotion in Man* (Springfield, Ill.: Charles C Thomas, Publisher, 1935).

APPENDIX A
Anatomic Movement Terms

JOINT ACTION

Joints are categorized according to the movement permitted. A *uniaxial* joint is said to have one degree of freedom, which means that movement can occur in only one plane. Typical of such joints are the knee, elbow, and fingers, which are hinge joints; and pivot joints, such as the two radio-ulnar joints of the forearm and the atlanto-axial joint of the upper neck.

Biaxial joints are those having two degrees of freedom wherein movement can take place in two planes. Condyloid and saddle joints are biaxial.

Triaxial ball-and-socket joints found in the shoulder and the hip allow movement in all three planes and are therefore said to have three degrees of freedom.

MOVEMENTS

All of the following definitions apply when the subject is standing in the anatomic position, that is, an erect stance with palms facing forward.

Flexion is the bending of a limb or segment. There is a reduction in the angle formed at the joint between the adjacent bones. From the anatomic position, flexion always occurs in the sagittal plane.

Extension is simply the return to the anatomic position following flexion.

Hyperextension is the continuation of extension beyond the normal anatomic position. It is not usually possible in hinge joints but is quite easy at the neck, lower spine, wrist, and shoulder. As with flexion and extension, it occurs in the sagittal plane.

Abduction is the lateral-plane movement of an arm or leg away from the midline of the body.

Adduction is a return to the anatomic position following abduction.

Hyperabduction is a continuation of an arm's abduction beyond a vertical position.

Hyperadduction is possible at either the hip or the shoulder and involves the movement of an arm or leg across the body toward the opposite side.

Circumduction occurs only at ball-and-socket joints. It is the cone-shaped movement that can be described by an arm or leg.

Inward or *medial rotation* takes place in the transverse plane around a longitudinal axis. The anterior humerus or femur rotates toward the trunk.

Outward or *lateral rotation* involves the rotation of the posterior humerus or femur toward the trunk.

Rotation refers to the turning of the head or trunk to the left or right.

A number of movements have names that specifically describe the motion of particular body parts. At the shoulder joint, the definition for flexion does not apply because an angle is not reduced. When the arm swings forward and upward, as in walking, it is said to be flexing at the shoulder. Extension is the downward swing in returning from flexion. When the arm moves horizontally forward from an abducted position, it is *horizontally flexing.* Returning the arm to the horizontally abducted position is known as *horizontal extension.*

At the ankle, rising onto the toes is termed *plantar flexion,* while bringing the foot up toward the shin is called *dorsiflexion.* Both are sagittal-plane movements. Standing on the outer edges of the feet is termed *inversion.* In *eversion,* a person would stand with the outer edges of the feet raised.

The forearm can *supinate* as the palm turns forward or upward. It is *pronated* when the palm is rotated downward or rearward.

The spine is said to *laterally flex* when one does a side bend in the lateral plane.

Elevation is a movement of the scapulae upward as in hunching the shoulders. *Depression* is a return from elevation. The scapula will *rotate upward* whenever the arm is raised and will rotate *downward* as the arm returns to anatomic position. When the arms are horizontally flexed, the scapulae will *abduct,* and of course, they will *adduct* or move closer to the spinal column when the arms horizontally extend.

APPENDIX B
Trigonometry Review

In dealing with right triangles, we know at the start that one of the angles equals 90° and that the sum of the other two angles must be 90° because in any triangle, all of the internal angles must total 180°.

TRIGONOMETRIC FUNCTIONS

The ratios of the sides of right triangles are trigonometric functions. When the angles are held constant, the sides will always be in the same proportion. In other words, regardless of the size of the right triangle, the ratio of the side lengths will remain the same if the angles are the same.

The function called the *sine* (abbreviated *sin*) of the angle is the ratio between the side opposite the angle and the hypotenuse, which is the side opposite the right angle and is always the longest of the sides.

The *cosine* (cos) reflects the relationship of the adjacent side to the hypotenuse. The *tangent* (tan) expresses the ratio between the opposite side and the adjacent side, and the *cotangent* (cot) refers to the ratio between the adjacent side and the side opposite the angle. In summary,

$$\sin = \frac{\text{opposite side}}{\text{hypotenuse}} \qquad \cos = \frac{\text{adjacent side}}{\text{hypotenuse}}$$

$$\tan = \frac{\text{opposite side}}{\text{adjacent side}} \qquad \cot = \frac{\text{adjacent side}}{\text{opposite side}}$$

In triangle ACB below, if we know one side and one angle other than the 90° angle, we can proceed to determine the other side, the other angle, or the length of the hypotenuse. Or if we know two sides, we can find the third

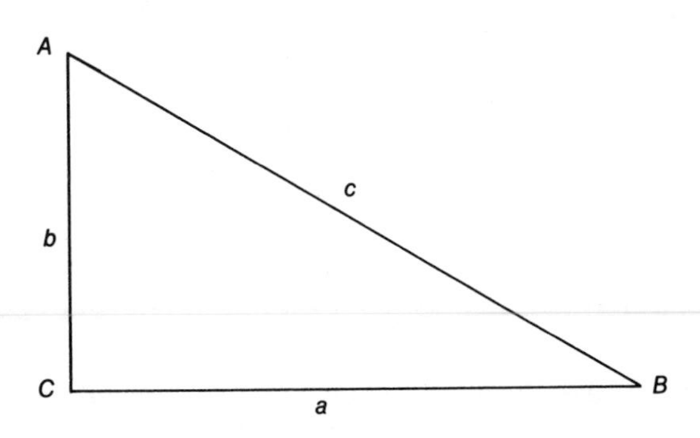

side by using the Pythagorean theorem. Or, knowing two sides, we can find the angles. In the triangle, capital letters A, B, and C refer to angles, while the sides are represented by lower-case letters a, b, and c. Thus,

$$\sin B = \frac{b}{c} \qquad \cos B = \frac{a}{c} \qquad \tan B = \frac{b}{a} \qquad \cot B = \frac{a}{b}$$

$$\sin A = \frac{a}{c} \qquad \cos A = \frac{b}{c} \qquad \tan A = \frac{a}{b} \qquad \cot A = \frac{b}{a}$$

Examples: Given side $a = 20$ and angle $A = 35°$, find side c.

Begin by locating an equation from the above group in which you can substitute both the known information and the unknown side. We can see that the only useful equation is

$$\sin A = \frac{a}{c}$$

because the other equations contain two unknowns.

$$\sin 35° = \frac{20}{c}$$

$$.5736 = \frac{20}{c}$$

$$c = 34.9$$

Given side $a = 20$ and side $c = 34.9$, find angle B. The equation of choice is

$$\cos B = \frac{a}{c}$$

$$\cos B = \frac{20}{34.9} = .5730$$

which is approximately the cosine of 55°.

Given side $a = 20$ and side $b = 28.6$, find angle B. Here we may use either the tangent or the cotangent.

$$\tan B = \frac{b}{a} = \frac{28.6}{20} = 1.43$$

which is approximately the tangent of 55°.

SQUARE ROOTS

Pocket calculators have nearly eliminated the need for slide rules and tables of square roots and of trigonometric functions. But it is handy to know how to extract a square root by paper and pencil. With a little practice, the procedure is easily learned.

To find the square root of 3214.02, begin by placing it under the radical sign and then pairing off the digits in both directions away from the decimal point. Place a decimal point above the line over the original decimal point.

```
     .
√ 32 14 . 02
```

Now determine the largest number which when squared will be equal to or less than the first pair of numbers (32). It appears that 5 is what we will use, and 5 will be entered directly above the pair while its square, 25, will be placed under the pair. As in long division, subtract 25 from 32 to get a remainder of 7.

```
   5  .
√ 3214.02
  25
   7
```

The next two steps require some care. The number above the radical sign (the quotient) is doubled and placed to the left of the most recent remainder, in this case the 7. The next pair of numbers is brought down and placed next to the 7 to make 714.

```
           5  .
        √ 3214.02
          25
10         714
```

The number 10 followed by a new digit to be added must be divisible into 714. The added digit is the maximum number of times that the newly formed number to the left will divide into 714. It appears that 6 will be close. Enter 6 in the quotient and to the right of the 10, making it 106. Six times 106 is 636, a number smaller than 714.

```
           5 6 .
        √ 3214.02
          25
106        714
           636
```

The remaining steps are repetitions of the earlier steps. The product 636 is subtracted from 714, leaving 78. The pair 02 is brought down and placed to the right of the 78. The 56 in the quotient is doubled and the 112 placed to the left of the 7802. To the right of the 112 is placed a new digit that will indicate the number of times the resulting number will divide into 7802. The process is repeated to the number of decimal places desired.

```
             5 6 . 6 9
          √ 3214.0200
            25
106          714
             636
1126          7802
              6756
11329         104600
              101961
```

The proof is found by squaring the quotient:

$$56.69^2 \approx 3214.02$$

APPENDIX C
Table of Trigonometric Functions

Degrees	Sines	Cosines	Tangents	Cotangents	
0	.0000	1.0000	.0000		90
1	.0175	.9998	.0175	57.290	89
2	.0349	.9994	.0349	28.636	88
3	.0523	.9986	.0524	19.081	87
4	.0698	.9976	.0699	14.301	86
5	.0872	.9962	.0875	11.430	85
6	.1045	.9945	.1051	9.5144	84
7	.1219	.9925	.1228	8.1443	83
8	.1392	.9903	.1405	7.1154	82
9	.1564	.9877	.1584	6.3138	81
10	.1736	.9848	.1763	5.6713	80
11	.1908	.9816	.1944	5.1446	79
12	.2079	.9781	.2126	4.7046	78
13	.2250	.9744	.2309	4.3315	77
14	.2419	.9703	.2493	4.0108	76
15	.2588	.9659	.2679	3.7321	75
16	.2756	.9613	.2867	3.4874	74
17	.2924	.9563	.3057	3.2709	73
18	.3090	.9511	.3249	3.0777	72
19	.3256	.9455	.3443	2.9042	71
20	.3420	.9397	.3640	2.7475	70
21	.3584	.9336	.3839	2.6051	69
22	.3746	.9272	.4040	2.4751	68
23	.3907	.9205	.4245	2.3559	67
24	.4067	.9135	.4452	2.2460	66
25	.4226	.9063	.4663	2.1445	65
26	.4384	.8988	.4877	2.0503	64
27	.4540	.8910	.5095	1.9626	63
28	.4695	.8829	.5317	1.8807	62
29	.4848	.8746	.5543	1.8040	61
30	.5000	.8660	.5774	1.7321	60
31	.5150	.8572	.6009	1.6643	59
32	.5299	.8480	.6249	1.6003	58
33	.5446	.8387	.6494	1.5399	57
34	.5592	.8290	.6745	1.4826	56
35	.5736	.8192	.7002	1.4281	55
36	.5878	.8090	.7265	1.3765	54

Degrees	*Sines*	*Cosines*	*Tangents*	*Cotangents*	
37	.6018	.7986	.7536	1.3270	53
38	.6157	.7880	.7813	1.2799	52
39	.6293	.7771	.8098	1.2349	51
40	.6428	.7660	.8391	1.1918	50
41	.6561	.7547	.8693	1.1504	49
42	.6691	.7431	.9004	1.1106	48
43	.6820	.7314	.9325	1.0724	47
44	.6947	.7193	.9657	1.0355	46
45	.7071	.7071	1.0000	1.0000	45
	Cosines	*Sines*	*Cotangents*	*Tangents*	*Degrees*

For angles greater than 45°, refer to the right-hand column and read upward from the column headings at the bottom.

APPENDIX D
Unit Conversions

Given	Sought	Multiply by
	Force (Weight)	
Pounds	newtons	4.448
Newtons	pounds	0.2248
Kilograms-force	newtons	9.8
Kilograms-force	pounds	2.2
Newtons	kilograms-force	0.102
	Mass	
Kilograms	slugs	0.06852
Slugs	kilograms	14.59
Pounds	slugs	0.031
Slugs	pounds	32
	Length (Distance)	
Meters	inches	39.37
Meters	feet	3.28
Feet	meters	0.3048
Yards	meters	0.9144
Meters	yards	1.094
Miles	feet	5280
Miles	kilometers	1.609
Kilometers	miles	0.6214
Kilometers	feet	3281
Inches	centimeters	2.54
Centimeters	inches	0.3937
Feet	centimeters	30.48
	Velocity	
mi/h	ft/s	1.467
ft/s	mi/h	0.682
mi/h	m/s	0.447
m/s	mi/h	2.237
mi/h	km/h	1.609
km/h	mi/h	0.6214
m/s	ft/s	3.28
ft/s	m/s	0.3048
km/h	ft/s	0.9113
ft/s	km/h	1.097
	Energy	
Foot-pounds	joules	1.356
Joules	foot-pounds	0.738

APPENDIX E
Cartesian Rectangular Coordinates

In a basic coordinate system, the x and y axes intersect at right angles and form four quadrants, which are numbered counterclockwise from the top right. The x axis to the right of the y axis is positive, while the x axis to the left is negative. The y axis above the x axis is positive, and the y axis below is negative.

An angle between 0° and 90° falls into the first quadrant; that is, its abscissa is to the right of the y axis and its ordinate is above the x axis. In this quadrant, both the sine and the cosine of the acute angle formed at the center are positive in sign.

For obtuse angles between 90° and 180°, the sine of the angle is equal to the sine of its supplementary angle and has a positive sign. The cosine is equal to the cosine of its supplement but has a negative sign.

The correct signs must be used when the sine law or the cosine law is used. We have not used angles greater than 180° in any problems in this text, but where they are encountered, the student needs to remember that in quadrant III both the sine and cosine are negative, and that in quadrant IV the sine is negative and the cosine is positive.

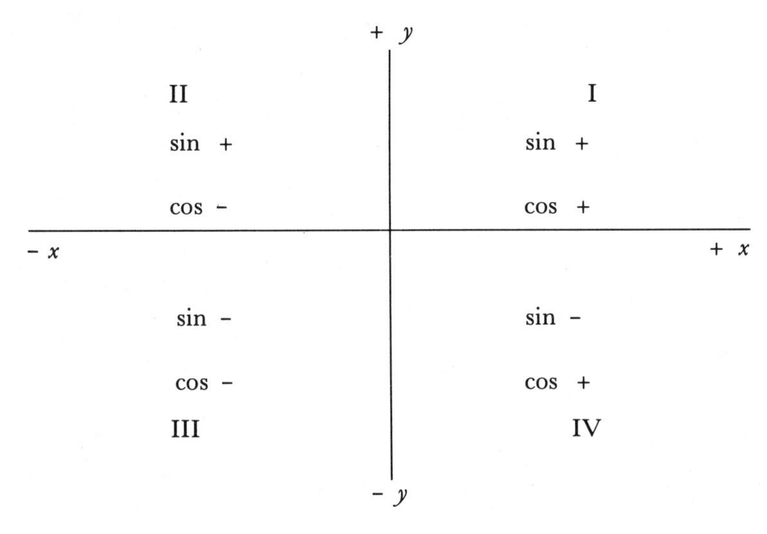

In summary,

Quadrant	*cosine*	*sine*	*tangent*
I	+	+	+
II	−	+	−
III	−	−	+
IV	+	−	−

APPENDIX F
Answers to Odd-Numbered Problems

Chapter 2

1. 73.3 ft/s
3. 11.4 mi NW
5. 31.7°
7. 28

Chapter 3

1. F_y is 52.48 lb, F_x is 60.38 lb
3. F_y is 12.5 lb
5. 30 lb

Chapter 4

1. 11.2 ft/s²
3. 5.5 s; 242 ft
5. 90 lb
7. $\bar{v}$ is 12 ft/s, v_f is 24 ft/s, a is 16 ft/s²
9. 683 ft/s (465 mi/h)
11. 25 ft/s
13. (*a*) 2 ft/s², –4 ft/s² (*b*) $\bar{v}$ is 15.7 ft/s (*c*) d is 550 ft

Chapter 5

1. 15 lb•ft
3. 251.2 rad
5. 9 rad/s

Chapter 6

1. 20 N
3. 480 J (480 N•m)
5. 650 lb
7. 4.7 ft

Index